STILL IN MOVEMENT

STILL IN MOVEMENT

Shakespeare on Screen

LORNE M. BUCHMAN

New York   Oxford
OXFORD UNIVERSITY PRESS
1991

Oxford University Press

Oxford   New York   Toronto
Delhi   Bombay   Calcutta   Madras   Karachi
Petaling Jaya   Singapore   Hong Kong   Tokyo
Nairobi   Dar es Salaam   Cape Town
Melbourne   Auckland

and associated companies in
Berlin   Ibadan

Copyright © 1991 by Lorne M. Buchman

Published by Oxford University Press, Inc.,
200 Madison Avenue, New York, New York 10016

Oxford is a registered trademark of Oxford University Press

Library of Congress Cataloging-in-Publication Data

Buchman, Lorne Michael.
Still in movement: Shakespeare on screen / Lorne M. Buchman.
p.   cm. Includes bibliographical references and index.
ISBN 0-19-506541-7
1. Shakespeare, William, 1564–1616—Film and video adaptations.
I. Title   PR3093.B8   1991
822.3'3—dc20   90-39821

2 4 6 8 9 7 5 3 1

Printed in the United States of America

To my parents and

to my teacher Ellie Prosser

Acknowledgments

My sincere thanks go to many friends and colleagues who have given generous assistance in the preparation of this book. I am particularly grateful for the guidance and care of Professor Charles R. Lyons of Stanford University and for the helpful suggestions and sensitive reading of Professor Barbara Hodgdon of Drake University. Gratitude for help with this book also extends to Larry Ryan, Marvin Rosenberg, Dunbar Ogden, William I. Oliver, Henry May, and Warren Travis—all of whom gave their ideas and encouragement in the process of completing the project. Special thanks to Marni Wood for never tiring of writing letters for research grants, and for doing all she could as chairperson of our department to promote the book. The contribution of Travis Bogard is gratefully acknowledged, not only for his attentive eye to the manuscript, but also for his thoughtfulness in introducing me and my book to the patient Sheldon Meyer of Oxford University Press. Recognition goes to Bernice Kliman and Kenneth Rothwell; they are at the center of research on Shakespeare on film in this country, and I value their collegiality. My students at Berkeley are my most exciting teachers, and I thank them for their enthusiasm and constant challenges. I am indebted to Bertrand Augst, Hugh Richmond, and Mark Collopy for their help in acquiring the films for study, and to Kim Salyer of Video Arts in San Francisco for his help with stills.

The book would not have been possible, nor so pleasurable to work on, without the film resources and staff of the Folger Shakespeare Library in Washington, D.C. My appreciation goes as well to the Mrs. Giles Whiting Foundation for the generous Humanities Fellowship in 1983–84, and to the University of California, Berkeley, for the many and various research grants and fellowships awarded over the last six years. I am grateful to Cambridge University Press for permission to reprint the chapter on Orson Welles's *Othello*, which first appeared as an article in *Shakespeare Survey*, Vol. 39, 1987.

To find words to express my thanks to David McCandless, Steve Vineberg, Frank Murray, Benjy Rubin, Ron Davies, and Daniel Goldblatt is a task more difficult than the book itself. My gratitude to my parents and family is also indescribable—they are forever reassuring and devoted. The selfless friendship and unswerving love of my wife Michele were with me in the writing of every word of this book; and, though he came around later in the process, and though he probably does not know it yet, my little boy Zachary has always been an inspiration.

Finally, to my great teacher and friend Ellie Prosser, I am forever indebted. She read and reread everything, tirelessly challenged muddy arguments, cleaned up the prose and graceless phrasing, and demonstrated all the while what makes her the finest of teachers: her ability to listen, to question, and to inspire with the gift of her abundant knowledge.

Berkeley, California   Lorne M. Buchman
July, 1990

Contents

STILL IN MOVEMENT

Through the Machine

I

The idea of Shakespeare on film conjures, for me, that vivid image in Chaplin's *Modern Times* when the tramp, diligently working on the assembly line, becomes so flustered by the increasing speed of the conveyor belt that, in earnest pursuit, he ends up in the great modern machine itself, processed through its gears and wheels. After four centuries of stage productions of Shakespeare's work, a new machine takes hold of the plays and processes them according to its own particular laws. Chaplin's point in his film has to do not so much with the wonders of technology, as with the relationship of the human being to the technological world. Similarly, in this study, I am primarily concerned with the new relationship between Shakespearean drama and the cinema, between the plays and the new medium in which they find realization. My approach focuses on performance, not on stage but on screen, an approach that proceeds from the assumption that the study of Shakespeare on film is a study of the plays as a function of cinematic space and time.

Although, in a few instances, a contrast of stage and film is illuminating to the questions I address, I am not, in any essential way, offering a comparison of the two media. The study is not concerned with judgements on which is the "better" medium for Shakespeare or the appropriateness of film for the playwright's work. Much of what the film achieves, the stage can duplicate or approximate according to its own principles, and vice versa. In what follows, considerations of stage practice often and inevitably arise, but they do so only to clarify how the drama becomes a product of the cinematic medium and how film activates the imagination of the spectator in a unique manner. This is a study that looks at the experience of seeing Shakespeare's plays in a relatively new medium, an experience never assumed to be at the expense of stage production or in a contest with it.

In addition, the process of adapting Shakespeare's work to the screen leads to productions that illuminate certain aspects of the plays and neglect others, often in ways more radical than productions for stage. There is, in any case, no such thing as "doing Shakespeare straight," and it is the essence of performance that new actors, directors, and designers will always bring to the drama insights, ideas, and ideological interests pertinent to the given historical moment. I do not mean to suggest that "anything goes." There are many ways to distort the plays and to use them for one's own peculiar cause, often with no relevance to anything in the text or, at least, nothing major in it. Similarly, it is a great bore to see "original" concepts that, in the final analysis, have only a tangential relationship to the play. All too often we witness productions with some striking new concept that, after the thrill of the first five minutes, wears itself out. Still, one cannot direct Shakespeare's plays (or any play for that matter) without a concept, without something to say. But that statement must come from careful thought and deliberation by all the artists involved, both individually and collectively. Performance relies on rigorous interpretation.

Granted, I chose many of the films under investigation in this study partly because of the compelling interpretive strategies of the filmmakers involved. In almost every case, the films offer penetrating insights into the plays and can stand with pivotal moments in their individual stage histories. But even more important than their interpretive approach is the filmmakers' exploitation of the medium itself, the way they utilize cinematic technique to realize the plays and to stimulate a distinctive imaginative response to the drama. Grigory Kozintsev, for example, almost entirely removes the Christian context of Shakespeare's *Hamlet* to make a point—through techniques uniquely filmic—about the political nature of the events in the drama; while I do not ignore the consequences of his approach, I also make no judgements about his interpretive accuracy. I am not concerned with whether the Soviet director's approach is "faithful" to the text or whether one can justify the cuts and emendations he makes. While I hope my discussion of Kozintsev's work betrays the respect and admiration I have for him and for his approach to Shakespeare, I do not base my analysis on the value or appropriateness of his conceptual focus.

II

In the last ten years or so, we have witnessed an explosion of scholarship in the area of Shakespeare on film. After the pioneering studies of Robert Ball, Roger Manvell, and Jack Jorgens (as well as Charles Eckert's collection of essays)[1] and the articles that appeared from the 'fifties onward in Shakespeare journals and their counterparts in film, the study of the playwright's work on screen continues to fascinate anyone involved in the teaching or study of dramatic literature and cinema. Shakespeare on film is now a regular part of the seminar offerings at major Shakespeare conferences; full-length studies

emerge with surprising regularity;[2] the University of Vermont publishes a
newsletter semi-annually;[3] collections of essays are growing in number;[4] and
relevant theater, film, and Shakespeare journals increasingly publish articles
on the topic. The most significant reasons for this development are, I believe,
threefold. First, given the growing interest in performance-centered criticism, film adaptations allow for many screenings and careful study and
provide an opportunity for many people in disparate parts of the world to
view the same production. Second, the importance of the cinematic medium
as a force in our culture cannot be underestimated. I am speaking not only of
film's popularity, but of recent developments in theory as well. Film theory
has become a sophisticated and penetrating discourse on contemporary aesthetic and cultural values, and theater scholars would do well to find nourishment from studies of cinema as they have for centuries from those with a
literary focus.[5] Shakespeare on film provides a specific and concrete example
of how central works in the history of Western culture take form by way of a
modern technological apparatus. And if, as Heidegger points out, "technology is a mode of revealing," a specific way of "ordering" the world, then it is
pertinent to ask how this process develops in the relationship of cinema and
Shakespearean drama. "Unlocking, transforming, storing, distributing, and
switching about are ways of revealing."[6] These are, in fact, the processes of
reorganizing the plays (what the filmmakers do), and are the issues to address
when asking how the play functions as a product of this technological medium. Finally, I have found that one of the most compelling aspects of study-
ing Shakespeare on film stems from the opportunity it affords to dance with
such partners as Orson Welles, Grigory Kozintsev, Peter Brook, Roman
Polanski, and Laurence Olivier. Surely, the proliferation of material in this
area is a result of the opportunity to learn from these figures, and I am sure
that many have found the same kind of stimulation from long hours with
them that I have.

Despite the amount and range of scholarship, however, the structure of
the studies written to date is largely uniform: the general pattern takes the
form of individual essays either on a single film or on a number of adaptations of the same play. My approach in this study is, for the most part,
concerned not so much with the individual films, or with comparisons of
different cinematic treatments of a given film (though I do, at points, structure analysis in both these ways), as much as with central issues that illuminate how the plays are operating as products of cinematic technique. The
central question of this work has to do with the way the films organize the
material of Shakespeare's plays to activate a particular imaginative response.
I have, therefore, isolated moments that clarify this process according to the
spatio-temporal structure of the medium and limit my analysis to directors
whose work provides the most fertile ground for the exploration of these

issues. Each chapter covers a specific topic and addresses questions pertinent to that topic, not through a comprehensive look at the films, but
through a detailed treatment of key sequences. I examine such issues as the
multiple perspectives through which we experience a given event on film,
the dynamic of key facets of *mise-en-scène*, the unique way the medium
contextualizes the action of the play, the spatial field of the close-up, and the
temporal structure that film gives to Shakespeare's work. In every case, I
am concerned with the new relationships among elements of the Shakespeare play that the films create, a concern that forms the basis of my
concluding chapter. I return most often to moments in the films of
Kozintsev and Welles; interestingly, these men have proven to be my best
teachers, and thus many of my examples come from their work.

III

Scholars of Shakespeare on screen tell, in detail, the various background
stories on the making of the films, and I send the reader to entries in my
bibliography for this information.[7] Still, in the following chapters I proceed
with the assumption that the reader has some basic knowledge of the films,
and it is important to review, briefly, the more pertinent facts.

Orson Welles made three major Shakespeare films, all in black and white,
and all starring Welles himself in the lead roles: *Macbeth* (1948),[8] *Othello*
(1952),[9] and *Chimes at Midnight* (released as *Falstaff* in the United States,
1966).[10] Welles developed the *Macbeth* film from two stage productions he
had previously directed: the Federal Theatre Project's "Voodoo *Macbeth*" of
1936 and a production for the 1947 Utah Centennial Festival in Salt Lake
City. He shortened the play by two-thirds, cut entire scenes, excised characters, added others, rearranged what remained of the play, and made the film
in twenty-three days. Welles has called his *Macbeth* a "violently sketched
charcoal drawing of a great play."[11] This "sketch," partly because of its
simplicity, and partly because of Welles's innovative filmmaking, provides a
chance to begin to address the primary issue of multiple perspectives in
cinematic adaptations of the plays.

Welles's *Othello*, on the other hand (a film he took four years to complete),
is one of the most beautiful and moving Shakespeare films to date. Although
free of the many production problems caused by the time constraints in
making *Macbeth*, *Othello* still gives evidence of Welles's special blend of
artistic brilliance and technical negligence. He shot most of it in Morocco
(except for the sequences detailing the material in Act 1, which he shot
mainly on location in Venice) in the west coast town of Mogador, and ably
used a castle there built by the Portuguese in the sixteenth century.[12] Although Welles won the Grand Prix for the film in the 1952 Cannes Film

Festival, *Othello* has received the least critical attention, and, unfortunately, of his three Shakespeare films, it continues to be one of the most rarely seen of all cinematic adaptations of the plays.

In 1965, Welles released *Chimes at Midnight*, his final Shakespeare film.[13] Like *Macbeth*, *Chimes at Midnight* has its origin in a stage production, this time directed by Welles and John Houseman for the Theatre Guild in 1938. Entitled *Five Kings*, the stage version was based on a collage of scenes from Shakespeare's two tetralogies—*Richard II*, the two parts of *Henry IV*, *Henry V*, the three parts of *Henry VI*, and *Richard III*. Casting himself in the role of Falstaff, Welles attempted to stage the entire story of England from 1377 to 1485 in one gigantic extravaganza; but the show was, according to most reports, a disaster.[14] Alternatively, Welles chose to base the film almost entirely on the two parts of *Henry IV*, with isolated moments from *Richard II* and *Henry V* (and a snippet from *The Merry Wives of Windsor*). With this material, and with a central focus on Falstaff and his relationship with Hal, Welles uses the cinema to create a penetrating examination of one of Shakespeare's most popular figures. The abundance of prose in the dialogue lends itself well to the cinema, and the dynamic possible with a collage of scenes from both parts of *Henry IV* opens the way for a rich exploration of Shakespeare's second tetralogy. Welles uses his medium in *Chimes* to explore the dynamic in Shakespeare's history plays of court and tavern, of "holiday" and "everyday," of political responsibility and misrule, and of a relationship between a royal prince and a tavern king. Most significantly, as a study of the complexities of the Falstaff character, the film illuminates how the Knight is emblematic of the contradictions, ambiguities, and tensions of the world Shakespeare presents in both parts of *Henry IV*—contradictions the film accommodates in new ways.

Chimes was shot in Spain. The Boar's Head set was located in the basement of a block of workers' flats in Madrid; the Gadshill robbery was shot in the Caso del Campo park of the same city; and the coronation of Henry V was set in the magnificent Cordova Cathedral. Outdoor scenes were set in the medieval villages of Castile, villages that resemble their English counterparts of that period.[15] In this film, Welles took on the responsibilities himself of producing, directing, scripting, and acting, as well as designing sets and costumes. As a result, *Chimes* has a unity and consistency of style that his other Shakespeare films sometimes lack.

Soviet director Grigory Kozintsev is both a sensitive artist of the cinema and a first-rate Shakespearean critic. His two Shakespeare films (both in black and white), *Hamlet* (1964)[16] and *King Lear* (1970),[17] are striking for the force of their visual power and for the boldness of their design. Kozintsev also wrote two books on Shakespeare: *Shakespeare: Time and Conscience*[18] and *King Lear: The Space of Tragedy*.[19] His first book includes chapters on his critical approach to Shakespeare, essays on *King Lear, Hamlet,* and the two parts of *Henry IV,* plus a diary based on the making of *Hamlet.* The diary is of special importance for this study because one can examine some of his thought processes as he put the film together. The second book is chiefly a diary of the making of *Lear* and voices, not only the struggle that went into its creation, but the many insights into the play that came out of the process as well. Kozintsev writes movingly and with passion. He communicates his insights on Shakespeare with an obvious love of the plays, a keen critical eye, and a compelling directorial imagination.

Kozintsev is a student of the great Soviet stage theorist and director Vsevolod Meyerhold. Though Meyerhold himself was only tangentially involved in the cinema, he had a strong influence on Sergei Eisenstein (who formulated much of his film theory on the basis of Meyerhold's theoretical principles of the stage), and, in turn, on Kozintsev himself. Throughout his writing, Kozintsev expresses his gratitude for the guidance of these two great artists. Kozintsev's link with Meyerhold and Eisenstein is most apparent in the manner by which form is integrally linked to the conceptual statement of his work. The precise rhythmic structure of his films, the meticulous physical work of his actors, the use of montage as his primary organizational strategy ("the complex unity of an image is revealed only in the clash of its various features")[20]—all give evidence of the influence of these men. Kozintsev also takes seriously Meyerhold's insistence that a director work freely with any given text to create a production pertinent to one's time: "A play is simply the excuse for the revelation of its theme on the level at which that revelation may appear vital today."[21] Kozintsev follows Meyerhold's advice in his Shakespeare films. He insists that we must see in Shakespeare, not irrelevant struggles of a past, but the vivid realities of the present: "the generations who admire classic art find in it not the fossil of an ideal, but the interests and feelings of their own time."[22] Moreover, each director must find in Shakespeare what is personally relevant; only then will the variety and multiple dimensions of the playwright's work attain meaning in time:

> [Shakespeare] belongs to every man. If, moreover, the man is an artist, he must find his own meaning for Shakespeare's poetry. And if all an artist can see in the famous plays is what has already been done with them, should he bother to undertake a production?[23]

Like Jan Kott, the Russian director's interest in the playwright is an interest in a contemporary:

> Shakespeare's words still ring, resonant and full. His verse is still blistering its hands today. . . . His plays seem to be written by someone close to us, by a man of our time.[24]

Perhaps the most famous of all Shakespeare films are those made by Laurence Olivier. Olivier directed and starred in three major cinematic adaptations of Shakespeare's work: *Henry V* (1944; in color),[25] *Hamlet* (1948; in black and white),[26] and *Richard III* (1956; in color).[27] His work has received much critical attention, which ranges from reviews of his performances and direction to detailed examinations of his concepts and textual work.[28] *Henry V* will always be remembered for the great Globe fantasia that frames the film, for the *mise-en-scène* based on medieval illustrations from the *Calendar of the Book of Hours* of the Duke of Berry, for its patriotic subtext (the historical significance of Agincourt in England in World War II), and for Olivier's rousing speeches as Henry V. His *Hamlet* is striking for its realization of Dover Wilson's speculations on the play, for its Freudian interpretation based largely on the work of Ernest Jones (since Olivier's film, stage productions repeatedly interpret the scene in Gertrude's "closet" as a scene in her "bedroom"), for the excising of Rosencrantz and Guildenstern and Fortinbras (which also meant the cutting of the "all occasions" soliloquy), and for the blonde Hamlet Olivier made famous. Finally, *Richard III* achieves its fame through the brilliant performance of the central figure (especially with respect to his physicality and spellbinding soliloquies), and the marriage of the great performers Olivier and Richard in an astonishingly powerful relationship. Add to this core performance the work of John Gielgud as Clarence, Ralph Richardson as Buckingham, and Claire Bloom as Lady Anne, and Olivier's film, in addition to its innovations and interpretive strategy, is significant for the series of great performances it records.

Two other films that receive attention in the study are Peter Brook's *King Lear* (1970; black and white),[29] and Roman Polanski's *Macbeth* (1971; color).[30] Both films are significant as statements about the unrelenting pain of psychological affliction and as examinations of the unmitigated violence that human beings, in quest for power, are capable of unleashing. Both directors use the cinema to realize the uprooting of human values—Brook by creating a bleak and attenuated universe in which the action unfolds (a universe not unlike the one Beckett creates in *Endgame*),[31] and Polanski by emphasizing the violence and bloodshed that comes from the hero's mad drive for power, a violence born of the tortured psyche of Macbeth himself (violence that obviously had tremendous significance for Polanski as a result of the Sharon Tate murder). While I do not focus on these films with the kind of detail I devote to Welles, Kozintsev, and Olivier, they are still pertinent for certain techniques that help one learn about how the plays function as products of the cinematic medium. Finally, in my chapter on the close-up ("Expanding Secrets"), I explore very briefly a scene from Tony Richardson's *Hamlet* (1969; color),[32] a film shot almost exclusively in close range and therefore

significant to the question of how the drama unfolds in this tight spatial
field.

IV

As a result of the kinds of questions central to this study, I choose to isolate,
in what is finally an artificial construct designed for the purposes of analysis, the temporal and spatial attributes of the cinematic medium. Time and
space work in the performing arts as a continuum, and for every instance of
a new spatial dynamic, there is, of necessity, an assumption of the implicit
temporal structure operative, and vice versa. It is therefore difficult to
isolate one from the other. How can one speak of a dynamic of inside and
outside spaces, of theatrical and filmic space, of multiple perspectives (all
topics I cover in what follows), without understanding how these features of
cinematic space operate in time? In a sense, therefore, the workings of the
temporal field are always an implicit part of the analysis of the spatial field.
Similarly, we "read" and experience time in film by the spatial landmarks
that unfold in its course. Time has spatial attributes. Though on one level
spurious, my isolation of time and space is meant only to clarify, in a
concrete way, how the plays function in the performance context of film. In
the act of viewing itself, such distinctions are moot.

In addition, I give more attention in this study to issues of cinematic
space than I do to those of cinematic time only because spatial innovations
of the medium are more pronounced in Shakespeare films than temporal
ones are. This is not to suggest that the temporal field of cinema is any less
significant to the central question of how the plays function according to the
particular nature of the medium. It is to say that, relatively speaking, filmmakers tend to adjust Shakespeare's own (already cinematic) temporal structure less radically than that of the spatial field in which the events unfold.
As a result, my overall analysis tends to focus on issues of space more than it
does on those of time. In the chapters on time, moreover, I am able to
explore, through Orson Welles's *Othello*, not only how the play operates as a
function of cinematic time and space, but also how the temporal structure of
the film illuminates Shakespeare's own exploration of time in the tragedy.

While I do isolate questions of time and space in the study, I do not
separate for analysis issues of the visual and aural fields. What one hears and
what one sees in film are so integrally linked that it would be absurd for a
book that assesses how Shakespeare films activate a unique imaginative
response to miss, or only to imply, this crucial relationship. That the visuals
of the films receive the most attention is a consequence both of the approach
that I adopt as well as of the filmmakers' own creative emphasis. Much of
what one can learn about Shakespeare on film is fundamentally connected

to the visuals it offers; the aural field is clearly of great significance (especially in juxtaposition with the visuals), but the changes in Shakespeare production that come from cinematic adaptations are far more a part of what Walter Benjamin calls the "new field of perception" (what he takes to be primarily a new *visual* field) of film.[33]

I return to Chaplin's image in *Modern Times* to ask a final question. Part of the power of the image of the human being and the machine lies in the shock of seeing those two elements juxtaposed in that particular way, the shock of a new relationship. Indeed, Chaplin's image is a potent illustration, in pictures, of the concept of defamiliarization (or of Brecht's *Verfremdungseffekt*) championed by the Russian Formalists in the early part of this century. Independently, both the human being and the machine are familiar objects; even a man working at a machine is a familiar sight. The picture of a man processed through the gears of a machine, however, creates a visual poetry that functions precisely because of the new relationship it opens, a relationship that sheds new light on the individual and the technical environment as well. Could this same process of revitalizing the familiar occur in the specific instance of a Shakespeare play viewed in the strange context of the cinema?

CHAPTER 1

Spatial Multiplicity: Patterns of Viewing in Cinematic Space

I

The spatial field of the cinema is one of multiple perspectives. In the context of film, Shakespeare's plays are products of a dynamic of camera angles, cuts, the proximity and distance of shots—all set in juxtaposition to the material of the aural field. The world unfolds in the cinema as a world of infinite relationships, of images, sounds, textures, and colors in constant motion, of elements joining in endless permutations. In the performing space of cinema, there are, as Wylie Sypher argues in a comparison of film and the cubist artifact, no "simple" locations;[1] it is a space in which dramatic action finds realization through a spectacle of multiplicity.

The universe of the film, however, finds completion only at the moment when a spectator is present, only when the observer works with the infinite relationships in the creative act of viewing. The film activates the imagination of the spectator. To borrow the phenomenological model of reading offered by Wolfgang Iser (which he formulates on the basis of Roman Ingarden's work), the Shakespeare film exists in a "virtual dimension"; "this virtual dimension is not the text itself, nor is it the imagination of the reader: it is the coming together of text and imagination."[2] Iser's model springs from the premise that reading is a process that operates on a program of continual modification as the reader anticipates, remembers, reconstructs, shapes, and reshapes various moments and events of the text while reading. Reading is a process locatable in this "virtual dimension," this unwritten part of text, this place of creative "filling-in," where past, present, and future commingle in an imaginative dance of anticipation and retrospection.[3] The substance of the written text, therefore, is not found in the object itself, but in the subject–object relationship, in a convergence of elements that makes precise description impossible.

12

How can one begin to explore the character of the "virtual dimension" in filmed Shakespeare? What, specifically, constitutes the dynamic interaction of filmic text and viewer imagination? One fundamental relationship in the imaginative journey of viewing Shakespeare on film (and one that results from the cinema's spatial field of multiple perspectives) consists of an interplay between moments of identification and moments of alienation. Directors play with multiple perspectives in their films to keep the imagination alive, to keep it off balance, as it were, never allowing the spectator to rest with any one point of view. The dynamic they create, in turn, makes us aware of the alternating sensation we experience as audience between involvement and retreat, between a subjective positioning and a more objective and contemplative point of view. Indeed, cinema has the capacity to give the spectator the opportunity to share the subjective perceptions and experiences of a given character, an opportunity made palpable because of the technical resources of the medium. Film exposes the world from the point of view of another person, it allows us to see *with* another. But the film can easily alienate us from that subjective sharing with a quick cut to a shot from another point of view. The shifts in perspective that film offers the spectator of Shakespeare's drama, together with the imaginative positioning that results, is the issue of primary concern in this chapter, an issue that informs, in varying degrees, the central arguments of this study as a whole.

I should stress at this juncture that in the process of experiencing alienation from an object we paradoxically see that object with even greater clarity. This principle is at the heart of the Russian Formalist notion of "defamiliarization": the familiar can appear to us with new precision when we have the luxury of a new point of view, a point of view born of myriad formal devices in art and literature. In that process we not only "see" the object again (rather than simply "recognize" it), but achieve a new understanding of the very mode of perception, the *way* of seeing, operative before the moment of alienation.[4] On the other hand, the term "identification" should not be equated with the common notion of "sympathy" for a character that critics have traditionally, and often incorrectly, built on Aristotle's arguments in the *Poetics* (though such sympathy is sometimes the result).[5] I mean instead a sense of being able to experience with the character a specific perspective, or attitude, through the resources of the film.[6] In addition, identification must be understood in relationship to the process of alienation that causes us to be aware of the identifying perspective in the first place. The reverse is true as well: we are aware of "being alienated" only in contrast with moments of identification. The paradigmatic relationship of identification and alienation is a reciprocal one.

I begin a study of how the spectator works with the multiple perspectives of cinematic space in terms of the identification–alienation dynamic by looking at an example from Orson Welles's *Macbeth*.[7] For Welles, Shakespeare's play is about the evil forces of nature "plotting against Christian law and order." Macbeth is a victim—the "ambitious man" whom the witches use to achieve their ends. (These quotations are taken from a prologue to the film that Welles eventually cut.) The voodoo image that Welles employs crystallizes his idea. In the opening of the film, the Weird Sisters form a clay figure of Macbeth, a voodoo doll in their hands. When the witches name him Thane of Cawdor, they place an amulet around the neck of the figure; when they prophesy that he will be "king hereafter," they place the spiky crown of Duncan on its head; when Macduff finally kills Macbeth, Welles intersplices a sudden match cut of the witches beheading the figure. Welles's visuals are unambiguous—Macbeth is at the mercy of supernatural powers of evil. In opposition to a critical tradition that stresses the importance of Macbeth's choice, Welles's film denies any place for the hero's free will. Like Jan Kott, Welles sees a Macbeth who comes to realize that "every choice is absurd, or rather, that there is no choice."[8]

At a pivotal moment before Macbeth kills Duncan, the hero speaks of an imaginary dagger in his "heat-oppressed brain." In the film, as we observe Macbeth struggle with this "dagger of the mind," Welles briefly interrupts our viewing with quick shots of the witches flashing a dagger before the eyes of the voodoo figure.

> Is this a dagger which I see before me,
> The handle toward my hand? Come, let me clutch thee.
> I have thee not, and yet I see thee still.
> Art thou not, fatal vision, sensible
> To feeling as to sight? Or art thou but
> A dagger of the mind, a false creation,
> Proceeding from the heat-oppressed brain?
> I see thee yet, in form as palpable
> As this which now I draw.
> Thou marshall'st me the way that I was going,
> And such an instrument I was to use.
> Mine eyes are made the fools o' th' other senses,
> Or else worth all the rest. I see thee still,
> And on thy blade and dudgeon gouts of blood,
> Which was not so before. There's no such thing.
> (2.1.34–48)[9]

The first issue to arise from Welles's treatment of the soliloquy concerns his choice to present it in voice-over. With this technique, the aural and visual fields of the moment occupy distinct and separate "spaces" in the viewing process. We hear a man's words but do not see him speaking. We simply observe him "in thought." Moreover, though we hear his words, and though we see the pain he registers on his face at this moment, we do not actually see the dagger. Unlike the private voice "made public" on the soundtrack, the perception of the imaginary dagger, though as much a part of Macbeth's consciousness as his intimate thoughts, is something the audience never shares.

To clarify this point, it is instructive to make a quick comparison between Welles's technique and traditional stage practice. The voice-over, as a filmic convention, grants us a special privilege—a chance to react to the hero in a way more intimate than the relatively distant experience of a soliloquy on stage. A spoken soliloquy in the space of a live performance is a public event, even though convention suggests the opposite. We witness a character speaking in a public forum, yet learn through convention to "read" the moment as an expression of private thought. The voice-over of film, though it, too, makes public someone's intimate struggles, works differently. Without direct speech, the soliloquy imitates the process in human experience of listening to the voices of one's own consciousness. The film convention takes us inside the character in a manner similar to the way we go "inside" ourselves. Phenomenologically, the voice-over offers an aesthetic experience that reproduces processes of self-perception that exist in the everyday world. At the same time, however, the experience is made even more complex in Welles's film because, in the visual field, the spectator simultaneously assumes a relatively objective, external point of view as he or she observes Macbeth in "contemplative action." Visually the audience is "outside," while aurally they are "inside." In addition, Welles chooses, in spite of the unique technical resources of his medium, not to realize Macbeth's vision; as with any stage presentation of this play, we in the audience accept the dagger as Macbeth's hallucination, as a phenomenon of his subjective experience, as an illusion we cannot share. Thus, though we are invited to identify with Macbeth through the aural field, we are simultaneously alienated from that identification because of our failure to *see* the imaginary weapon he speaks of. What we hear and what we see in the film collide.

Welles does not stop with this simple division of the aural and visual in the performance space he builds. He makes the viewing process more intricate as the soliloquy progresses by creating a tension specific to alternating visual perspectives. As we hear "Is this a dagger which I see before me," the camera zooms in for a close shot of Macbeth's face, bringing us into a visually intimate relationship with him, albeit from the "outside." On the words "let me clutch thee," Macbeth tries to grasp the dagger, clutching furiously at airy nothingness. Just as he makes that gesture, the picture blurs for a moment; it comes back into focus as his arm retreats. Welles restores the spectator's perspective with a close-up of Macbeth's face. Then, on the words "thou marshall'st me where I was going" (transposed from a later point in the soliloquy), we actually share Macbeth's perspective as he looks towards the chambers where Duncan sleeps; as we share this point of view the picture again goes in and out of focus. We are now "inside" Macbeth. The speech continues, "Art thou not, fatal vision, sensible," and we are looking at Macbeth again in full-frame close-up. He makes a small turn away from the camera on "dagger of the mind," and, within this tight visual field, what would be almost imperceptible on stage or in the frame of the long shot here resonates with much power; he has pulled away from us; he is distant. Within the confined space of the close-up, every blink of the eye is a powerful gesture. Welles thus alienates the viewer into an awareness of his or her distance from Macbeth with a subtle but effective sign. On the final line of the soliloquy in the film ("there's no such thing"), the shot blurs for the last time and then becomes clear in a regular medium shot of Macbeth.

With the techniques he employs in the visuals of this moment in *Macbeth*, Welles activates a powerful dynamic. We identify with the hero as we share his experience. This happens in two distinct ways. First, and most obviously, we share Macbeth's perspective when he looks to Duncan's chambers. The spectator sees what the hero sees and assumes his point of view. We also share Macbeth's experience in a second, subtler way: as the picture goes in and out of focus we participate in the waves of Macbeth's own imagination. We find ourselves on a visual journey guided by a fluctuating and troubled consciousness. All this is contrasted with shots in which we assume the normal, objective point of view of the camera when we observe Macbeth from the outside. The relationship crystallizes at the moment Macbeth tries to grasp the dagger: we identify with him as we experience the disoriented blur of his consciousness, yet, in the same instant, experience alienation as we observe a man clutching at air. Welles creates a dialectic unique to cinematic presentation. That he also intersplices shots of the witches flashing the dagger before the eyes of the clay figure alienates us as well. We perceive this moment of Shakespeare's play from several perspectives, and as we participate in this process, we come to understand part of what constitutes the spatial field of filmed Shakespeare.

Olivier's treatment of the opening scenes of his *Richard III* provides a felicitous comparison with Welles's direction of the dagger soliloquy, a comparison that furthers our understanding of the imaginative dynamic of alienation and identification. These opening scenes (from Edward's coronation through the wooing of Anne) are governed by the machinations of Richard and offer a compelling juxtaposition of dramaturgical content and cinematic form. Olivier's use of the camera—and hence our perspective on the events—is at the center of his directorial scheme; as Richard manipulates those around him, he also manipulates the perspective of the spectator.

Olivier the filmmaker grants Olivier the performer (playing the role of Richard, the consummate actor) control of the camera. He uses cinematic technique to give form to Shakespeare's own dramaturgical design, and parallels in his camera work the dazzling schemes of one of the playwright's most potent vice-figures.

In his thorough study of the evolution of the medieval vice, Bernard Spivack sees Richard and Iago as the Shakespearean characters who represent the figure in its maturest form.[10] Indeed, as Spivack's work makes evident, the vice is (and he and his cousins always have been) one of the most theatrically compelling figures in the history of art. Evil works on stage. The figures of vice always make their counterparts of virtue pale by comparison. Shakespeare's vice-figures Iago and Richard delight their audiences by their sheer brilliance, by their knowledge of human need (and their concomitant ability to manipulate, knowing what those needs are), by their theatrical craft (staging events to conform to the script they write), by their wit, by their humor, by the intrigue of their scheming. What marks the Shakespearean vice and gives him distinction, however, is the manner by which he, in all his machinations and self-serving designs, proceeds with a wink to his audience. Richard III enacts his brutality with a theater audience of conspirators. Shakespeare's vice-figure talks to his audience, lets them know his plans, asserts his motive, takes them into his confidence, and communicates, in the subtext of his actions and words, that everyone besides himself and his conspiring audience is stupid and deserves what he or she gets. Watching Richard on stage or screen we are suddenly aware of our own tacit participation in his actions. Not that we condone evil (or the destruction caused by the vice); we simply open ourselves to a figure of charisma and perverse charm and find ourselves compelled to take the journey he offers.

The delight in the vice-figure's actions, therefore, unfolds in a constant tension with the spectator's own moral standards. Built into the reception of a play like *Richard III* are the conflicting processes of associating with the vice as a conspirator and of feeling repelled by the immorality of the actions we tacitly agree to encourage. Olivier's cinematic treatment gives form to this very conflict. In the film Richard is the consummate actor and playwright, managing his rise to power with flawless ability to dissemble. Olivier carefully creates a *mise-en-scène* that reflects as much about the location of the action as it does about the nature of that world as a habitat for Richard to bustle in. The feel and look of the set is theatrical. With the colorful decor, large open spaces, and painted backdrops, there is an artificiality about the world the drama unfolds in, something reminiscent of the theater itself. In his *mise-en-scène*, Olivier thus reinforces the notion of Richard's writing out the text of events the audience witnesses. The hero becomes a palimpsest of

roles that the spectator deciphers. Implied in Richard's activities are the
filmmaking strategies of Olivier, made that much more complicated by the
presence of the director himself as actor performing the role. Vice-figure
maneuvering, film directing, and playwriting—all converge in a fascinating
design of parallel manipulations. Shakespeare, Olivier, and Richard III
coalesce.

Within the space of this theatrical world, Olivier offers his audience
distance through medium and long shots of the central figure in the opening
soliloquy, which he deftly contrasts with his use of close-up and traveling
camera.[11] From his opening "Now is the winter of our discontent," to "Was
ever woman in such humor wooed," Richard appears, in the context of the
multiple perspectives of cinematic space, now as a distant schemer, now as a
conspiring friend. As he looks directly into the camera in close-up, one feels
as if Richard is speaking directly to each individual; he offers a secret that
empowers the spectator in an act of confidence. At one moment Richard
will loom large in close-up on the screen, giving visual form to his overwhelming presence and to the persuasive rhetoric he speaks. But a long shot
will show Richard in the distance, small, unreachable, suddenly foreign. In
retrospection we see that the long shot provides a new perspective on the
powerful close-up that preceded it. It gives us an imaginative breath of fresh
air, as it were, and consequently we are alienated from what we have now
come to learn was a surrender to Richard's spell while the close-up was
operative. In other words, the two contrasting shots produce a spatial dynamic that parallels Shakespeare's dramaturgical design: with the close-up
we associate with Richard as we participate and delight in his scheming,
only to become alienated from that involvement through a perspective that
reminds us of our moral distance from, while paradoxically highlighting our
susceptibility to, the vice. The spectator does not rest with one perspective
or the other; the process works in a dynamic relationship between points of
view.[12]

But where Welles employs the techniques of editing to create the multiple
perspectives on the dagger soliloquy, Olivier, by contrast, has Richard himself flagrantly control the perspectives we have on him. Olivier "blocks"
himself in the opening moments of the film; each time, therefore, that we
achieve a new point of view it is because vice-figure (who is also actor and
filmmaker) has openly and obviously made that choice for us. For example,
the opening soliloquy begins in a long shot of a distant Richard, who, with a
few steps, approaches the camera on his opening words to place him in
medium range. On "Our bruised arms hung up for monuments," Richard
gestures to the wall to display the very artifacts he speaks of, and he asserts,
for the first time in the film, his position as host and architect of this
theatrical event. Similarly, Richard offers "To the lascivious pleasing of a

The opening soliloquy of Laurence Olivier's *Richard III*. The spatial dynamic of the long and close shots activates viewing patterns that parallel our fascination with and alienation from the vice-figure's scheming.

lute" as if it were a cue to some stagehand operating the music; on the soundtrack we hear the instrument he speaks of. Four lines later ("I, that am curtailed of this fair proportion"), he walks away from the camera, turns back to look at us momentarily on "Deform'd, unfinish'd, sent before my time," pivots, walks away again on "Why love foreswore me in my mother's womb,"[13] and delivers the next ten lines in long shot. Then, in a move of brilliant theatrical (Richard) and cinematic (Olivier) manipulation, the vice-figure approaches the camera (on "Then, since this earth affords no joy to me") to place himself in full close-up. He looks directly at us for a moment, makes a quarter-turn and, in a profile shot (looking back at us over his shoulder, never taking his eyes off the camera), he begins to walk. But the camera is no longer stationary; this time it moves with him. The effect is that Richard pulls us along, that he makes us move where he wants us to move, that he has us on the theatrical tether that Shakespeare gives his irresistible vice-figures. In the film Richard holds an invisible string to the camera: the eye through which the spectator observes the world. Hence, Richard controls our perspective. He pulls us along and we are suddenly conscious that we want to move with him, that it is exciting to follow his splendid intrigue, that we are, in fact, conspirators (and it is no accident that the traveling shot begins on "I'll make my heaven to dream upon the crown"). Olivier gives spatial form, unique to the resources of the cinema, to this essential element of Shakespeare's own design. Perspective moves with Richard.

As the soliloquy continues, Olivier stops the traveling shot in favor of a long shot of Richard on a rostrum, which he then contrasts with a close-up that culminates, appropriately, on the line "And *frame* my face to all occasions" (italics mine). This last close-up shot only reinforces, through a lovely pun of spoken and visual language, the relationship between what Richard wants to create and what the spectator sees. It is important to stress that the traveling camera and close-up work only in a dynamic relationship with the long shot. Or, to put it another way, Richard (Olivier, Shakespeare) teaches us that one gains power (theatrical and otherwise), not through flagrant control and imposition, but through a program of careful contrivance that seemingly gives the victim-spectator the room to make independent choices. The traveling shot works because Olivier first gives the viewer the comfort of the long shot. Richard's murderous plans work because he smiles while he plays the villain. Shakespeare's theater works because his vice-figure, though morally reprehensible, is so intelligent and so compelling a character.

II

The above examples of Richard and Macbeth provide an opportunity to explore the nature of the spectator's response to an individual character's soliloquy in the context of cinematic space. The direct speech of Olivier's delivery contrasts with Welles's voice-over technique, in that the former presents a public, theatrical figure who serves as our host and as the chief architect of the events we witness, whereas the latter depicts the introspective musings of a tortured man. Though the two soliloquies have obviously different dramaturgical purposes, they both activate the imagination of the spectator in a paradigmatic interplay of identification and alienation. In one film the villain controls our perspective; in the other we move with the blurring waves of the central figure's troubled soul. The patterns of viewing become more complex, however, when other characters enter the scene and the number of permutations of perspectives increases. In the banquet scene of the third act of *Macbeth*, Welles offers another moment that builds on the paradigm.

The scene begins solemnly. In a long shot of the entire table, Macbeth toasts Banquo as all present raise their goblets. On the words "would he were here," the camera zooms in on Macbeth's face, wide-eyed and terror-stricken, supposedly registering his first sight of the ghost (though at this point the spectator cannot see his vision). The confused guests lower their goblets. We watch the stunned Macbeth as the shadows of those goblets pass over his face. Then the camera cuts to Lady Macbeth, and the per-turbed expression on her face communicates clearly her concern that her husband will reveal all. Another shot of the confused guests is followed by a long shot of Banquo's empty chair. The camera then cuts to Macbeth, who points to the chair. In a moment of suspense, Welles slowly pans the point-ing finger and the huge shadow it projects on the wall. Suddenly, we share Macbeth's vision and see an empty table—except for the ghost of Banquo sitting at the far end. A reaction shot of the terrified Macbeth follows. The director cuts back to the ghost, now with blood running down its face. The next shot shows Lady Macbeth staring at her husband anxiously. We see her look at Banquo's chair. But this time, when we share her perspective, the cut to the end of the table shows an empty place. As the scene progresses, we see that, within the multiplicity of cinematic space, the ghost is there for Macbeth but not there for all other members of the feast. The pattern then continues: a shot of Macbeth will be followed by a shot of an empty table with only the ghost at the other end; alternatively, a shot of Lady Macbeth or the guests will be followed by a shot of a full table and Banquo's empty place.

A stage director of *Macbeth* must make a choice for the banquet scene; the moment can clearly work with or without the ghost of Banquo, as the stage history of the play suggests.[14] There is, however, a trade-off evident with each choice. The 1623 Folio includes a direction for the ghost's entrance, and to bring the ghost onstage is to offer the spectator the opportunity to see what Macbeth sees and to participate in his experience. But the moment can also work simply with an empty chair, because Shakespeare puts the focus in this scene on the reactions of Macbeth and his guests and not on any specific activity of Banquo's spirit. Unlike the Ghost in *Hamlet*, the phan-tom of the banquet scene does not speak, does not appear to anyone but the hero, does not claim to come from purgatory or for revenge. It exists for Macbeth, and only Macbeth. To share the perspective of the guests and fix our eyes on vacancy leads us to conclude, with them, that Macbeth has gone mad. To see Macbeth respond with alarm to an empty chair is horrifying for an audience and serves as a reminder of the hero's descent into illusion. Each alternative creates a different emphasis, weighs dramatic irony on one side or the other, and stresses one aspect of the complex dynamic of the scene.

Whatever the trade-off on stage, the potentials of cinematic space offer a third possibility. Through cutting, Welles's film realizes the ambiguous status of the ghost—it is and is not there, simultaneously. In other words, Welles uses the resources of his medium to reveal, as part of the landscape of this performance, the disparity between Macbeth's subjective vision and the perceptions of those around him. Of course, the conventions of live performance can communicate this ambiguity as well; we believe, when watching this moment in the theater, that Banquo's ghost, whether present onstage or not, is a vision unique to Macbeth. But in the film the spectator is able to observe both perspectives in a new relationship. The perspectives of cinematic space create a very specific kind of tension, one built on a dynamic that suspends the spectator in a collision of identification and alienation. Unlike the dagger soliloquy, however, the matrix of this scene is more complex. The viewer experiences the process of involvement and distancing not only with respect to Macbeth's point of view but with respect to those around him as well; consequently, the imaginative interplay of alienation and identification operates in a more elaborate and intricate pattern.

In *Chimes at Midnight*, Welles works with multiple perspectives in a similar way, creating a tension in the imagination of the spectator that sustains a fundamental ambiguity present throughout the entire Falstaff story on film. In the rejection scene at the end of the film (*2 Henry IV*, 5.5), Welles reinforces a feeling of ambivalence about the new King's act by creating a dynamic of identification and alienation from the perspectives of both Hal (now King Henry V) and Sir John. Welles visualizes the coronation of Henry V elaborately. The King rides in grandeur on his horse through a crowd of cheering subjects. The sheer delight of the crowd is in direct contrast to the rigid and lifeless soldiers in Henry IV's court evident throughout the film (a contrast that clearly indicates one side of Welles's strategy in his valorization of Henry V). Following this scene is a dissolve to the arrival at Westminster Abbey and to a relatively quiet, more somber procession within the walls of the Cathedral. Intermittent shots show Falstaff (Welles's other major point of focus) making his way boisterously through the crowd to find the King. The hundreds of extras in Welles's film, the ornate and overwhelming Cordova Cathedral, the elaborate costumes and decorations of royalty—all combine to legitimate the coronation (and the power structure it signifies) by a dazzling theatricality.

In the midst of all the formality, a voice cries out, breaking the order of the ceremony: "God save thee, my sweet boy! My King! My Jove! I speak to thee, my heart!" Falstaff falls on his knees, but Hal remains with his back turned—we share the moment, in a low-angle shot of the King, with Falstaff. The cathedral goes silent. Without turning around the new King speaks, "I know thee not, old man. Fall to thy prayers." On his next line ("How ill white hairs become a fool and jester!"), Hal turns to face his erstwhile friend. This low-angle shot of the King in his finery, seen from Falstaff's perspective, gives evidence of his tremendous power; Hal carries gracefully his flowing robe, scepter, and crown. The image of him at this moment, surrounded by the spears of soldiers and the high, majestic vaults of the cathedral, clearly reveals at least an ostensible difference from the youthful figure of the tavern; it is an image that epitomizes what Stephen Greenblatt sees as the "illusion of magnificence" of Renaissance royal power. But, in the film, something continually interrupts what this same critic convincingly argues is the spectator's role in creating the ideal ruler, whom we "deck out" through our imaginary forces.[15] As the King proceeds with his rejection speech, Welles films it from the alternating perspectives of Falstaff and Hal. Now we assume the King's position with a high-angle shot of the bewildered Falstaff; now we take the Knight's perspective with a low-angle shot of this strikingly majestic figure. The director continues this pattern to the end of the scene, ensuring that our sympathies are displaced in a pattern of identification and alienation.

The moment as it is filmed maintains the tension between our responses to both Falstaff and Hal. The shots of Falstaff from Hal's point of view fill us with compassion as Falstaff's aged features, half-smiling eyes and slightly quivering lip strike us with the kind of pain critics repeatedly note in this scene.[16] On the other hand, the youth and strong presence of the new King, standing amidst the grandeur of Westminster, is a necessary if not altogether welcome element, vital to the restoration of what we have come to see, in the film at least, as a war-torn and diseased nation.[17] From the low-angle position we witness the might of Hal's majesty, recognize that he is not, in fact, "the thing [he] was," and achieve a certain respect for the evolution and growth of the Prince to this point.[18] Yet, as Welles films the moment, we cannot help but experience rejection with Falstaff and feel his pain intimately. Welles alienates us from identifying with Falstaff, however, not because we lose our sense of the Knight's pain, which is at the center of the film, but because Hal's new position from above, his power, his maturity, his strength, are all vital to the healing of a diseased nation. Moreover, we identify with the Knight as we observe the King in a position of power, a position Falstaff wishes to achieve but will never gain; with Hal's perspective we observe the Knight and identify with the new king's need to solidify

The rejection scene of Orson Welles's *Chimes at Midnight*. Welles creates a tension and sustains ambiguity by way of the alternating perspectives of Falstaff and King Henry V.

the roles (one above and the other below) of this relationship, a solidification that ultimately breaks them apart.[19]

One can imagine a director approaching this scene in a way that would substantiate the claim that "power thrives on vigilance."[20] Indeed, Falstaff's public call to be welcomed by the new king produces the opportunity for the kind of act of rejection that theatricalizes the King's behavior and re-inforces its "illusion of magnificence" for all spectators, both within the play and in the theater itself. In other words, the director makes the traditional statement that we can understand Hal's preparedness for kingship by his rejection of that which cannot be accommodated, at least ostensibly, by conventional images of royal power. Contrariwise (and as many would argue is the case with *Chimes at Midnight*), the director could weight the balance on the side of Falstaff and emphasize the pain an insensitive King inflicts on the lovable Knight. But Welles's film creates a deadlock and not a privileging of one side or the other. In one sense, the film allows us to participate in a very private exchange between these two men; it opens the possibility to see not only *through* their eyes, but to travel the intimate space *between* those eyes. And by putting the spectator in this otherwise exclusive space, Welles creates a location for performance impossible before the advent of cinema.

In his direction of the nunnery scene in *Hamlet*, Grigory Kozintsev similarly situates the spectator in a private space between two characters who are at a point of painful separation. Unlike Welles, however, who gave us no way to look at the moment outside of the perspectives of Hal and Falstaff, Kozintsev alienates us from the intimate exchange of Hamlet and Ophelia by giving the spectator the opportunity to share the vantage point of Claudius and Polonius, two important observers of the interaction. The King and his advisor go off to hide behind a curtain in the great hall of the castle, leaving Ophelia standing alone in a shaft of light that pours in from the stained-glass window above. Ophelia soon discovers the pensive Hamlet on a staircase; she stands on one side of the balustrade and Hamlet on the other. For Kozintsev, this encounter is an encounter of prisoners. Ophelia comes across as a bad actress in the scene, desperately trying to enact her father's commands, speaking the required words, returning the gifts, trying to mask her love for Hamlet; but every line and contour of her body, everything expressed in her eyes, the timber of her voice, the shy bow of her head—all contradict her words. Hamlet walks around the balustrade and seizes her arm violently. He knocks the token she holds from her hand, and its fall echoes in the empty hall, in the void and hollow world of Kozintsev's Elsinore. Then, in close-up, Hamlet holds Ophelia up to whisper in her ear, "I did love you once." The spectator receives a privileged perspective on the moment, and Kozintsev stimulates a sympathy for the young lovers by letting us in on what is secret. The next perspective we have, however, is from the viewpoint of Claudius and Polonius, a perspective that immediately alienates us from the private and intimate vantage we just shared. As the spectator shares the perceptions of Claudius and Polonius the meeting of Hamlet and Ophelia appears very different indeed; from their perspective we see only the violence of Hamlet's actions, the intensity of his anger, without the love that counters and makes more human his gestures in close-up. From behind the curtain, we see Ophelia acting out the script of her father, without the signals, so glaring in the space of the close-up, of her pain and her love for Hamlet.

The levels of identification and alienation are complex in this moment of Kozintsev's film. First, unlike in the rejection scene in *Chimes*, the contrast of perspectives is not one between individuals, but between two groups. Kozintsev sets in juxtaposition the perspectives of Hamlet and Ophelia on the one hand and that of Claudius and Polonius on the other. For the Russian director, as for Bertolt Brecht, "the smallest social unit is not the single person but two people," and dramatic action unfolds in a dynamic of human relationships.[21] The result of the cinema's capacity to give spatial form to this dynamic is that it offers the spectator a sense of two distinct forces, one behind the curtain and the other in the center of the great hall. By so doing, Kozintsev highlights his social and political focus in the film[22] and builds the scene in a fundamental tension between private and public space, between shattered personal dreams and political contrivance. And in the dynamic interplay of the film, the distant, public viewpoint sheds light on our intimate perspective on private need, exposing a powerful juxtaposition in Shakespeare's tragedy. Second, we learn through the course of this scene that we witness a level of interaction between Hamlet and Ophelia that is not for all observers, that we enter an exclusive space; we see something that Claudius and Polonius cannot. The result is that we are made aware of how Kozintsev empowers his viewing audience by making them less restricted observers compared to the others in the scene. We are thus conscious of our own role as spectators in the moment, and find ourselves the only witnesses of the love that remains between these two figures. Simultaneously, we are aware of the limited perspective of Claudius and Polonius as we share their position. Most important, however, is the way the filmmaker offers a multiplicity of perspectives on the moment: we see the scene develop from the perspectives of Hamlet and Ophelia, then shift to a close-up on their intimate exchange, then we look from behind the curtain—all the while conscious of our own place as onlookers. With every shot, the filmmaker alienates the spectator from one level of identification to create another level in a dynamic sequence.

III

The film medium, in addition to its capacity to offer the spectator the chance to share, within the spatial field, the perceptions of a given character, also provides the opportunity to participate in a given state of mind. Just as Welles uses the medium to articulate Macbeth's consciousness through the blurring of a shot, so, too, directors of Shakespearean films continually employ the cinema to realize subjective experience in a specific series of spatial relationships. In other words, the governing force that leads to identification is not so literally "point of view," but an individual character's subjective state. The filmmaker can then employ the spatial field to distance the viewer from that subjective state by creating other spatial relationships that stand in contrast to it. What one discovers, however, in examining how the identification-alienation paradigm operates in this case, is that what distances the spectator can, in and of itself, stimulate a new level of subjective involvement. To put it another way, different devices designed to offer identification with different aspects of subjectivity work to alienate and comment on each other. I return to Orson Welles's *Macbeth* to illustrate this point.

A pattern of alienation and identification broadens across Welles's entire film in his visual realization of Macbeth's nightmare in juxtaposition with images depicting his ambiguous stature (i.e., Macbeth as "huge" domineering tyrant, *vs.* Macbeth as "tiny," defeated human being). On the one hand, Welles uses the resources of film to invite the spectator to share, through a series of visual techniques, Macbeth's convoluted picture of the world. Indeed, the spectator experiences much of the tragedy through extreme high- and low-angle shots; the director thus activates the viewer's imagination to see the world in the same demented way, and from the same skewed perspective, as does the hero himself. Spatially, the castle is a labyrinthine structure carved out of solid rock that becomes more complicated and confusing as the film progresses. Welles shifts various pieces of the set throughout the film, and, with Macbeth, we find it difficult to know where we are at any given moment or what awaits us around the next corner. The world beyond the castle is equally confusing as long as we are with Macbeth. Now we are in an open field; now, by a cliff; now, in a void of darkness. In this film, as Jack Jorgens points out, "journeying seems an act of the mind—with a few steps Macbeth traverses the distance from his aspiring castle atop a mountain to the hill where the witches reside."[23] Through montage, camera angle, visual distortion, and a labyrinthine set, the film *Macbeth* realizes a descent into a chaotic universe by inviting the viewer to merge his or her experience of the action with the subjective experience of the hero.

Welles alienates our identification with Macbeth's experience, however, by creating a second series of visual statements about the ambiguous stature of the hero. Shakespeare's treatment of Macbeth demonstrates that he is at once a hero of grandeur and a small, lost figure. On the microcosmic plane he emerges as a tyrant who carves out his passage with his "brandish'd steel." But on the macrocosmic plane, he is but a speck of dust. The specta-

Orson Welles's *Macbeth*. Macbeth the "giant" towers over his tiny subjects.

tor, aware of this division in the character Macbeth, must accept the ambiguity implicit in Shakespeare's hero. As Angus points out in the fifth act:

> Now does he feel his title
> Hang loose about him, like a giant's robe
> Upon a dwarfish thief.
>
> (5.2.20–22)

Indeed, Macbeth is both the "giant" and the "dwarf," the monster and the
conscience-haunted, ignominious man. The imagery Shakespeare employs
to articulate this ambiguity has led Caroline Spurgeon to remark,

> The imaginative picture of a small, ignoble man encumbered and degraded by
> garments unsuited to him, should be put against the view held by some crit-

ics . . . of the likeness between Macbeth and Milton's Satan in grandeur and sublimity.[24]

Welles parallels Shakespeare's imagery by using his camera to portray Macbeth as both great and small. In Act 1, scene 7, for example, when Lady Macbeth goads her husband to murder Duncan, Welles foregrounds the hero in low- and high-angle shots that exaggerate his size. The film-maker makes a visual statement that corresponds to Macbeth's and Lady Macbeth's emphasis on what it is to be a "man"—a definition of manhood that, like the distorted size of the hero, is bestial and beyond human propor-tion. Similarly, in Act 3, scene 1, Macbeth towers over Banquo in a low-angle shot as he assumes his new position of power. Later we see Macbeth on his throne for the first time and, through deep-focus photography and another low-angle shot, he appears a giant surrounded by tiny, obsequious subjects. When the murderers enter to plot Banquo's death, they appear very small next to the domineering King. At this moment, Welles uses a high-angle shot from behind the King's shoulder that dwarfs the murderers as they stand before Macbeth.

But Macbeth does not remain the towering giant for long. After the banquet scene, he decides to seek out the witches for more information. He leaves his castle and goes to the heath where he first met the Weird Sisters. There a storm begins and Macbeth is left alone, contending with the fretful elements. Thunder and lightning accompany his call to the witches as two sere trees and a Celtic cross flash in silhouette behind him. On the heath, Macbeth appears windblown, alone and small. Welles does not bring the witches on for this scene (we only hear their voices), and all we see is a hero surrounded by blackness. As we hear the first "apparition" speak ("Beware Macduff") Macbeth appears as an infinitesimal spot on a black abyss. We witness "the single state of man" reduced to proportion—a tiny, poor player "strutting and fretting his hour upon the stage." Within the microcosm of Dunsinane, Macbeth is the "giant." In the context of the greater universe, however, he is the "dwarfish thief"—a pinspot in a universe of blackness.

Put together, the pattern of identification and alienation forms a complex imaginative dance in the juxtaposition of these two lines of Welles's visual poetry (i.e., ambiguous stature *vs.* demented psyche). On the one hand, the perception of Macbeth as dwarf and giant causes the spectator to stand "outside" the hero's subjective experience; we not only share the distortions of his troubled mind but find distance from that experience in the realiza-tion of how context determines his stature. The two filmic statements about the hero work in a dynamic. Yet the fluctuation of Macbeth's stature is also, arguably, a reflection of his own self-image. That is, we see Macbeth as the giant as long as he conceives of himself in those terms; we see him as the dwarf at crucial moments of vulnerability. In a sense, then, one recognizes that the visuals articulating Macbeth's stature are also part of what leads the viewer to identify with the hero's experience. Paradoxically, what alienates us from such identification are Welles's distorted pictures, his shadows, and his labyrinthine set; because from the outside we see Macbeth follow the illusory, we know he is enmeshed in a world of shadow. We learn that his self-aggrandizement is self-deception. These are the elements that counter the identification we achieve when, participating in the hero's own self-perception, we see him as great and small. The dynamic thus operates in both directions.

In Kozintsev's *King Lear*, the sequence following the love contest demonstrates, in a manner that has significance for the whole film, elements that operate simultaneously in a reciprocal pattern of identification and alienation. We are again dealing with the representation of a state of mind distanced by a statement of ambiguous stature, and vice versa. Just as Welles realizes the consciousness of his hero through his distorted visuals, so, too, does Kozintsev use his medium to articulate the rage of Lear after his experience with Cordelia and Kent early in the film. Kozintsev traces the King's departure from the throne room (the location for 1.1) with a rapidly moving, backward-traveling camera. Lear leaves the castle with determination as attendants repeatedly call out: "An hundred knights for the king!" "Saddle the horses!" He storms through the stables, often looking directly ahead, while he chooses the horses for his journey. He then enters the kennels to select his dogs, pointing to a Great Dane on the right, then to a wolfhound on the left. He gestures to birds of prey, eagles, falcons. As he moves he barely looks at the animals he chooses; he does not see; he is king only through the blind gestures of authority. Indeed, there is little difference between Lear's treatment of human beings and his treatment of the beasts—all are subject to his whims. The speed of the sequence, the chaotic feel stimulated by the tracking camera, the visuals of horses, huge dogs, wild birds, scrambling attendants—all combine to articulate the panic of a furious king. Kozintsev, like Welles, utilizes cinematic techniques to create a chaotic universe that causes the viewer to merge his or her experience of the action with the subjective experience of the hero. We experience the moment through the filter of Lear's unbridled anger.

Kozintsev alienates our identification with Lear's experience, however, precisely the way Welles does for his hero in *Macbeth*. The Russian director creates visuals to articulate an ambiguous stature. After choosing the animals that will accompany him on his journey, Lear climbs the stairs outside of the castle while masses of peasants prostrate themselves before him. Torches blaze along the walls. A crane shot reveals the hundreds of peasants kneeling before the mighty King, increasing the spectator's sense of Lear as the all-powerful giant. Indeed, Lear's ascent reflects his own self-image as one above humanity. When he reaches the top of the castle walls, we view the "giant" in close-up as he stands before his kingdom, renouncing his daughter. He shouts:

> The barbarous Scythian,
> Or he that makes his generation messes
> To gorge his appetite, shall to my bosom
> Be as well neighbor'd, pitied, and reliev'd,
> As thou my sometime daughter.
>
> (1.1.116–120)

We share Lear's perspective momentarily as we look at the sweep of humanity, in all its relative tininess, prostrated before him. Suddenly, however, Kozintsev challenges that perspective of the all-powerful "giant" with the next shot from the point of view of the peasants below. Lear no longer appears to us as the great and mighty ruler of this world, but, as we view him from below in a low-angle long shot, we perceive a man dwarfed by the huge, stony edifice on which he stands, insignificant and alone. Kozintsev parallels Shakespeare's design as his visuals reveal a king who towers over the world only in his own mind and whom we must understand as one whose greatness is equaled only by his nothingness. Like Macbeth, Lear is both great and small, and Kozintsev's exploitation of high- and low-angle shots, close-ups, and long shots combine to articulate this shifting stature.

One could argue that Lear's diminutive appearance is also what he feels himself to be at this moment and what he desperately tries to ignore. Indeed, we quickly learn in the following scenes at Goneril's castle of his growing self-doubt. Thus the "tiny" Lear is, I would argue, a critical element of the hero's experience at this moment; the wounds of the love contest are open and sore. His feelings of rejection are painful because he is a man who has the capacity to recognize his small, unaccommodated stature. The seeds are present from the very start. The picture of a small, "dwarfish thief" is thus a way of inviting the spectator to share an aspect of Lear's experience that is seemingly hidden. The lonely, pitiful figure that distances us from a sense of his rage and power, therefore, also works in a reverse pattern to articulate part of what is at the core of his being. In this inverted pattern, the moments suggesting power and kingly rage alienate the spectator's association with the vulnerable Lear. We thus face a figure both great and small, and move in and among the fluctuations of Lear's own changing self-image at a critical point in the tragedy.

The sequences we have covered in this chapter demonstrate how multiple perspectives are part of cinematic space and how the spectator works with that spatial field according to an interplay of identification and alienation. It is a paradigmatic relationship in the "virtual dimension" of Shakespeare on film that operates on varying levels of complexity. The film medium offers the spectator an opportunity to travel in a new space, to take a special journey through Shakespeare's drama (and, as such, it can defamiliarize us from old ways of seeing the plays). We travel the lines of tension and conflict in the plays, within the consciousness of individual characters as well as among many different figures, as the camera opens for us a new field of perception and new spatial relationships. The cinema activates the imagination in a manner different from that of the stage, even though one often discovers that, in their approach to the screen, filmmakers parallel what Shakespeare achieves for his stage. Of course, Shakespeare's drama itself activates patterns of identification and alienation; this does not, however, alter the central point that the cinema, through the perspectives it affords, stimulates a unique meeting of spectator and play.

What this investigation teaches as well is that filmed Shakespeare, and the cinema's unique way of performing the Shakespeare play, is accessible through an understanding of the patterns of response that the film triggers for the spectator. The specific paradigm of identification and alienation, in addition to illuminating how the Shakespeare film operates, gives evidence of the viewer implied or produced by the film. We can understand the multiple perspectives of the film only in the context of the observer who works with those perspectives. In other words, one begins to discover something about the subject constituted by the filmed version of a Shakespeare play, a subject that is not an independent essence, but a participating factor in a complex relationship.

CHAPTER 2

Inside-Out: Dynamics of *Mise-en-scène*

I

Gaston Bachelard writes of a "dialectics of outside and inside" when teaching about the scope of the poetic imagination and pointing out how one can learn about the structures of language (and the mind that works with that language) through spatial configurations.[1] Bachelard gives the poetic image a spatial vocabulary: he offers a way of charting the image in terms of spatial relationships. In phenomenological terms he helps us understand the attributes the mind gives to language and, as he proceeds, he communicates the dynamic behind that process. By exposing the dialectical field of the poetic image, Bachelard points out the dynamism behind that image, a dynamism one can grapple with in terms of spatial concepts. His discussion of dialectical constructs of inside and outside is particularly significant for the workings of the *mise-en-scène* of film and is pertinent to the way the choices of individual filmmakers generate meaning in adapting Shakespeare's own poetry and dramaturgical design. The following analysis springs from Bachelard's work.

The specific nature of the spatial field in cinema activates an imaginative response that seems to disturb what appears, on the surface (as photographic realism), to be a clear delineation of spatial realms. The cinema has the capacity to juxtapose inside and outside worlds and create visual images of opposition; it also has the capacity to undermine that seeming antithesis, to unsettle, ultimately, and reconcile what seems disparate. Film presents Shakespeare's world through a *mise-en-scène* that renders the vastness of intimate space as much as the intimacy of a vast exterior. The concern here is, not simply the contrast of close-up in broad space or the distant shot in the confinement of the frame (though both are important techniques), but the imaginative interweaving of open and closed performing spaces in the film and the dialectical activity that results from that interweaving.

The *mise-en-scène* of Kozintsev's *King Lear* operates, in part, through the dialectic of the inside and outside worlds, of the closed and the open, of the confined and the expansive. Though at first glance these worlds appear to exist antithetically, further study shows a far more complex interrelation. "At the slightest touch asymmetry appears."[2] The outside world is measureless, void of natural growth, rocky, and wasted: it is a world inhabited by the "bare, forked animal" that Lear discovers in the process of his journey out-of-doors. In the exterior landscape of the film, humanity emerges in its most "unaccommodated" state as hundreds of beggars and peasants wander this forsaken universe. The rough textures of their garments, the scrapes and calluses on their bodies, the sweat on their faces and the dirt in their pores form the visual details of an otherwise unforgivingly barren terrain. In the emptiness, the viewer discovers, as Lear eventually does, the habitat of "the thing itself." In terms of the dialectics of the *mise-en-scène*, this outside world forms one pole of the film's visual landscape.

Juxtaposed with this harsh and desolate exterior are the interior scenes depicting a world protected by and adorned with the accoutrements that shelter humanity in "civilized" society. In contrast to the rough and coarse-woven sackcloth and rags of the peasants outside, the dress of the courtiers is made of predominantly smooth, silken materials. Furs accent the costumes, "stressing not only the kinship of men and animals but the pampered isolation of the rich from the poor."[3] The textures contrast with those of the outside world, yet, simultaneously, connote the domestication and exploitation of the wild beyond the castle. The inside brings with it, despite its sense of confinement, a hint of the presence of something beyond, something borrowed to inform it and dress it up. From the beginning, Kozintsev sets up the relationship of inside and outside in terms Lear will later perceive as a relationship between those who are "sophisticated" and those who are not, between a world that owes the worm its silk; the beast, hide; the sheep, wool; the cat, perfume; and one that has no such debt.

The inside world restricts; it is claustrophobic and static and contrasts with the constant motion outside. Kozintsev fills both interior and exterior worlds with people; but while those outside are constantly wandering through the vast nothingness of their world, the figures of the inside are restrained and still. Suffering in life is everywhere in this film, but, ironically, there is hope only among the most wretched, only among those able to move beyond a static and stultifying court. Movement in the outside is a sign of faith and promise, and Kozintsev constructs a space that can accommodate that movement, a space that contrasts in the film with a hopeless and rigid interior. That movement, however, bears the weight of the greatest tragedy in the film: the pain of starvation, cold, loss, loneliness. It is a tragedy that we trace back to the inside world of politics, lies, and the indulgence of petty needs. The outside unveils what comes from policies inside that privilege the few and create the suffering of the many. That is part of the fascination of Kozintsev's *Lear:* the inside dictates the form of the outside landscape.

Kozintsev's depiction of Lear's kingdom gives evidence of the lies of the court, especially those of Lear himself. The King speaks of his land in terms of "shadowy forests," with "champains rich'd" and "plenteous rivers." Nothing could be more opposite to what Kozintsev chooses to show us in his film. For the Soviet director, Lear's kingdom is not rich and plenteous but a barren, empty wasteland. Lear's words of fertility and abundance belie the visuals of starkness and sterility. Kozintsev renders Lear's vision of his land, the space of his kingdom, another aspect of his blindness; he knows nothing about what really exists beyond his own sheltered environment. The map Lear uses in Act 1, scene 1, which he claims to be a sign of his bountiful kingdom is, in the film, a false representation of the true geography of Britain. Similarly, the words of Goneril and Regan used to express their "love" are, like the map, deceptive signs. Cordelia's "nothing," on the other hand, is, like the wasteland itself, the reality. Thus, the outside does find a place in the court (through words) at the moment Cordelia responds to the pleading of her father. Kozintsev joins the seemingly cold and callous response of Lear's youngest daughter (an activity on the inside) with a harsh exterior landscape indifferent to the sufferings of humankind. Shakespeare's language and Kozintsev's *mise-en-scène* combine to form a compelling relationship between human utterance and spatial context, between a direct and biting reply to an impossible request and a landscape that seems to deny the chance of human survival. As Lear misunderstands Cordelia, so, too, does he misunderstand the reality of a barren and deserted kingdom. His blindness is extensive.

Kozintsev constructs both interior and exterior spaces with elements that are moving and elements that are still. The first sequence of the film begins with a close-up of the stones and cracks of the dry and barren land of the kingdom. In the visual field, nothing could be more static. On the soundtrack, however, there is movement: we hear the slow and even footfalls of a single man walking off-screen. Soon his feet enter the frame—his old, worn shoes move over the rubble. The man holds a crooked walking stick that aids him in his steps. The image of wandering in nothingness permeates this film and is central to Kozintsev's spatial design: the "space" of tragedy is the "space" of human growth, of movement, of discovery, of pain. The Russian director uses, as an element that most precisely defines the outside space of his film, the image of wandering humanity. Eventually, even the king himself will wander as one among many who simply roam the static earth. The camera pans from the man's feet to focus on a wheelbarrow, inside of which a child sleeps, peaceful and still, as its father pushes it along (an image that encapsulates the relationship of movement and stasis). Subtly, Kozintsev increases the distance and width of his shot to reveal an entire line of peasants following in single file. They move among the sharp rocks of the land, rocks standing like tombstones in a lifeless desert that goes on forever; it is difficult to detect a horizon. Suddenly, there is a close-up on one of the peasants, who blows a long blast on an old and primitive horn. The cry of the horn echoes throughout the land, and that piercing sound, that reverberating call for community—later echoed by Lear's own howls—delineates the vast space of Kozintsev's world as much as the visuals themselves do. The audience imaginatively follows the invisible route of sound to chart a spatial field; the sound becomes a force of movement in a static land. Immediately following the blast of the horn, hundreds upon hundreds of peasants appear. In a completely attenuated universe, human beings increase by the moment, and, as they do, a sense of suffering increases proportionately. The stony, pockmarked faces of the ragged peasants resemble the landscape itself; the film matches the broad spaces of the long shot with the limited space of the close-up as the surfaces of earth and human being become one. We move with the peasants on our way to a huge stone castle now visible in the distance.

Following the short dialogue between Kent and Gloucester (1.1), the scene cuts to the inside of the castle, an inside world Kozintsev also constructs through a complex relationship of stasis and movement. The specific techniques, however, differ from those he uses for the outside. Once inside, we hear footfalls again; but, unlike the grinding of the peasants' feet on the stony ground outside, these paces echo through the castle halls with sharp clarity. The camera first picks up a young Cordelia, dressed in a light-colored gown, running down a staircase. She joins her sinister-looking older sisters, dressed in black, and the three walk into the throne room, the camera tracking backwards as they move. Goneril goes over to join Albany, with his men stationed behind him; Regan then finds Cornwall and his entourage. Cordelia goes to stand by a lady-in-waiting, glances lovingly at France, but is quickly made to stand soberly with the others after a subtle but stern nudge from her lady. Servants, attendants, nobility, royalty stand absolutely still and silent, waiting for the arrival of the King.

> Everything is firmly placed, immovable, and subordinated to an unchanging order. . . . They all stand stock still, as though eternally glued in place. . . . [Lear] is sure that the world is stable.[4]

The court at this moment in the film stimulates a recollection of the very first image of stasis in the outside world before the old man enters the frame; the people of the court are as still as the tomblike rocks in the lifeless desert outside.

But there is movement inside as well, movement that contradicts the order and ceremony that exists on the ostensible level of Shakespeare's play and exposes the seeds of chaos and conflict below the surface. While all remain in place, the camera—and hence our perspective—roams freely, and through a collage of faces the film offers the spectator a visual preamble to the conflicts that shortly will emerge. We see Edmund eye father and brother with a menacing glance. We are aware of the discord between Albany and Cornwall when the two men glare at one another in a quick sequence of close-up shots. The faces of Goneril and Regan suggest antagonism not only toward one another but toward Cordelia, who appears in the following shot. The entrance of France and Burgundy is the introduction of another conflict as both men vie for the hand of Cordelia. While the court may appear calm on the surface, this is, in reality, a highly charged moment with an extremely complex dynamic. As the camera moves in the stillness, our perspective follows the underlying lines of tension, and we witness the dynamics of Shakespearean drama in a newly defined space—the space of hidden conflict.

In the opening moments, what creates spatial definition in the inside world is not the actor but the movement of the camera in a static setting. We experience space in film through the journey of the camera, a journey (in the specific instance of a Shakespeare film) that follows the subtextual undercurrents of the moment. To put it another way, part of the spatial field of cinema comes from the capacity of the camera to realize a vertical meaning of the text, to travel and hence expose the level of dramatic action operating underneath, to give form to an aspect of the performance text that is "below." That movement of the camera, however, works only because it contrasts with all that is still. Motion and stasis define the spatial field of the scene.

Kozintsev thus reconciles inside and outside through the juxtaposition of movement and stasis while showing the two worlds to be antithetical. On the one hand the opposition of inside and outside is crucial as a way of documenting and tracing Lear's journey; on the other hand, the boundaries are not so precisely defined, and the two worlds operate as discursive elements that share key properties within the total *mise-en-scène*. Lear must go outside to learn of his life inside. The new world beyond shocks the King into a state in which he can listen to humanity with a different part of himself, in which he can care about another's comfort, in which he can ask forgiveness, and in which he ends his demands for self-gratification. Lear's experience thus parallels the interweaving of elements in the *mise-en-scène* itself. As we shall see in the following chapter on the storm scenes, Lear learns through the contrast of environment, through moving into a new and opposing spatial context; but what he learns is not unique to the outside,

nor is it only about his status outside. It is also, and perhaps most impor-
tantly, about the world within—within the court, within his past, and
within his heart. Lear is a bridge that links inside and outside, and his
journey, while fundamentally dependent on their opposition, reconciles the
two worlds.

Lear moves inside himself by moving to the outside world, both physically and psychologically, through his madness. That madness is a kind of
ecstatic displacement, a stepping outside of the self to spur a closer vision of
the inside, a journey given dramatic form through the dynamics of
Kozintsev's *mise-en-scène*. Bachelard writes on this very paradox.

> And if we want to determine man's being, we are never sure of being closer to
> ourselves if we "withdraw" into ourselves, if we move toward the center of the
> spiral; for often it is in the heart of being that being is errancy. Sometimes it is
> in being outside itself that being tests consistencies. Sometimes, too, it is closed
> in, as it were, on the outside.[5]

Lear's "errancy," both in the sense of the mistakes he makes as well as in the
wandering that follows upon those errors, is a movement outside of the self
given the dramatic form of madness, a specific transition in the film from
the stasis of the inside to the wandering movement outside. Inside and
outside thus form a dynamic in Kozintsev's *Lear*; we witness at once a
physical movement in the film from the inside to the outside, a psychological transition of a similar nature taking the form of madness, and a spiritual
journey of discovery that brings a man closer to himself in the act of standing outside. The inside and outside work together as opposed and reciprocal
agents.

II

Orson Welles's *Chimes at Midnight* creates a dialectic of inside and outside
spaces even more complicated than does Kozintsev's *Lear*, because the interior worlds of court and tavern themselves produce a dialectical relationship. In addition, Welles specifies the outside world of the film most power-
fully in the battle of Shrewsbury sequence, thus defining that exterior
through a critical event in the drama. That sequence is not, by any means,
the only use of outdoor space. There is as well the Gadshill robbery, the
rolling, snowy hills of Gloucestershire that Falstaff and Shallow travel, the
space outside the tavern—with the court in the background—and the
streets of the town. It is, however, the moment in the film that defines most
clearly the outside world of *Chimes* and specifies why the interior worlds
appear the way they do. The outside space of the Shrewsbury episode
works in the richest of relationships with the indoor worlds of the film, and

I isolate the battle to focus discussion on this aspect of the spatial world of *Chimes*.

Welles defines the inside world of *Chimes at Midnight* in a relationship of court and tavern, a relationship that at once shows a fundamental disparity as well as a certain affinity between them. The shots at court are static and controlled, emphasizing the stature of the King and the rather rigid, businesslike, "everyday" world of Henry's castle. The high ceilings and concrete walls of the castle give a sense of confinement, and the directed shafts of light coming from small windows above add to the sense of a stark and alienated political world. The tormented King dwells within his cold castle walls, and on more than one occasion we see him walk pensively by a line of soldiers, rigidly at attention with their spears. Welles uses low-angle shots to depict Henry on his throne upon a rostrum, visually emphasizing the loneliness and isolation of his power and position. Wellesian film critics often compare the court of Henry IV with "the cold vaults of the Thatcher Memorial Library" in *Citizen Kane*.[6] The comparison is appropriate.

The tavern world, on the other hand, is one of energy and celebration. In contrast to the static shots of the court, "every time the camera is in this setting, it is alive with the energy of festive motion."[7] Especially in the scenes taken from the first part of *Henry IV*, Welles fills the Boar's Head with a sense of life and holiday that is wholly absent from the cold and exacting world of the King's fortress. Dancing crowds, half-naked whores, laughter, and the music of the tavern replace the rigid soldiers at attention in court. We are, once again, working with a relationship of movement and stasis, this time between the two interior worlds of the film.

Like the inside and outside of Kozintsev's *Lear*, however, the opposition of court and tavern is only one level of the dynamic: Welles reconciles the two worlds as well, and, by so doing, allows his spatial design to articulate an important feature of Shakespeare's text. The link between court and tavern in this film is the fact that both worlds are cut off from the outside and, as such, are protective and sheltered environments.[8] With the exception of the few shafts of light coming from the high windows of the castle, both worlds are isolated from the sun. The narrow corridors of the tavern find their counterpart in the labyrinthine corridors of the court. Falstaff is king of the tavern world, sitting like royalty on a thronelike chair and ordering his cronies about in a way reminiscent of Henry himself. In short, both court and tavern are similar as isolated worlds unto themselves; they strike the spectator as independent microcosms built to shelter those within, containing parallel, though unallied, political organizations. Welles taps an important aspect of the two parts of *Henry IV* by showing that even though these two worlds are fundamentally opposed as "everyday" and "holiday," they are similar as cocoons into which the characters of the play seem to retreat from the pain and horror of rebellion and civil war; retreat, in short, from the world outside. Thus, the film illuminates an ambiguous relationship central to Shakespeare's design. Court and tavern are not simply antithetical worlds reinforcing the divisiveness of England; they are both environments in which giddy minds are consumed with the games of everyday politics or of holiday misrule. In both worlds one recognizes the tension "when thieves cannot be true one to another." Falstaff's charge refers as much to the thieves of Richard's throne, who have fallen into disputing factions, as to the Gadshill pranksters themselves. The two worlds, though antithetical, mirror one another.

The sequence of the film that reveals the profoundest reality of the *Chimes* world outside is the battle of Shrewsbury. It is a world of destruction, of death; a battleground for civil strife, a field in which honor is a value without precise definition or understanding. And like the two poles of the inside world, opposing factions make up the dynamic forces of the outside. In this world, however, we witness hand-to-hand combat; at Shrewsbury the forces are real human beings colliding in a battle that characterizes the world in terms of loss, suffering, and the incomprehensibility of conflict among the nobility. The acute exposure we have to every horrifying detail of this war and the capacity of the film to bring the spectator into the place of every soldier's fall, every blow that adds to the senseless bloodshed of man against man, is the product of an entirely new spatial field in the experience of Shakespeare's history play. Welles creates an outside that stands in bold contrast to the "protected" inside but that simultaneously serves as the field where all aspects of that interior world—of court and tavern politics—meet.

When Worcester returns from the King, silent about his enemy's offer for peaceful negotiation, we watch him ride on horseback in a stark, wind-blown plain as Welles shoots him approaching the rebel camp through a maze of spears held by soldiers. The spectator sees the world through the artifacts of war. The instruments of death are everywhere. Upon hearing Worcester's news, Hotspur calls his army to battle. Immediately Welles cuts to Falstaff, who confidentially explains to Hal that honor is a "mere scutcheon." The juxtaposition of the rebels and the Knight at this moment is an important prelude to the battle itself, because Welles ultimately demonstrates that Falstaff's critique of honor, on one level at least, makes a great deal of sense indeed. *Chimes at Midnight* does not simplify and reduce Shakespeare by adopting the Knight's point of view on honor. The film does show, however, that Falstaff's word suggests an important corrective to the attitude of warmongers like Worcester.

Following Falstaff's lecture to Hal, we see soldiers lowering knights in armor from trees on to their horses. We then hear the regular beating of war drums along with trumpet calls to battle. As the war machine is set in motion, Welles focuses on elements of the battle slowly gaining momentum. The horses on both sides begin to move through the fog. Infantry advance slowly. Gradually everything seems to speed up, the horses galloping while the shrieks of battle cries echo through the land. The two sides charge one another. Meanwhile, the image of Falstaff on foot, as he "leads" his men to battle, strikes a note of absurdity in an otherwise sober moment in the film; he stands still, waving his soldiers on with his sword, his arm circling like a windmill. As all the soldiers run by him, Falstaff goes off to hide behind a tree.

When the encounter finally begins, we witness in great detail the suffering of men in war. Welles shoots close-ups of the fall of each man and makes immediate every stab, every punch, every horse's tumble. We are made to listen to the groans and shrieks of the fallen soldiers and, simultaneously, to the barbaric cries of attack around them. We hear a cacophony of clanging metal when the instruments of war meet in combat. We see the brutality of man against man as the camera moves as fast as the battle itself, at one moment revealing a soldier's final gasp for life and at the next showing an aerial shot of the entire bloody scene. The war in *Chimes at Midnight* is not an abstract event viewed from a single point of view, or realized through the metonymy of, say, an alarum, as it is onstage. In Welles's battle of Shrewsbury the audience find themselves in the center of the event, sharing the perspective of the terrified soldiers. As the battle ends (the sequence takes six or seven minutes to play), Welles produces a horrifying picture of the entire scene in silhouette against the grey sky. Soldiers struggle for life in the slimy mud of Shrewsbury. Commenting on the battle, Welles said,

> we shot with a big crane very low to the ground, moving as fast as it could move against the action. What I was planning to do—and did—was to intercut the shots in which the action was contrary, so that every cut seemed to be a blow, a counter-blow, a blow received, a blow returned.[9]

In the translation from play script to cinema, Welles transforms the battle of Shrewsbury by transforming the spatial field of action to produce a distinct world "outside." He uses the unique resources of film to define a terrifying space of human violence. Shakespeare relies on a few conventions in writing the battle for the stage. He uses the aural convention of trumpet sounds, the excursion across the stage of men in arms, the specific combat of certain key figures, the use of various characters as messengers of the details of the battle, and quick snippets of action to produce the sense of varied activity. On the other hand, Welles exploits the visual and aural resources of his medium to realize the battle: the "noblemen," in close-up, who lie "stark and stiff / Under the hoofs of vaunting enemies" constitute

the outside of *Chimes*. It is—in vivid images—a world "where stain'd nobility lies trodden on, / And rebels' arms triumph in massacres." A close-up on
the dying soldier looms on the screen as large as the battlefield itself, and his
sufferings become defined in an entirely new spatial context; the consequences of war, the deaths of human beings "outside," unfold in a dynamic
with inside scenes in which scheming politicians plan battles that murder
the innocent. In the film, the specter of scenes of collusion like, for example,
the one among Northumberland, Worcester, and Hotspur (*1 Henry IV*,
1.3.131–302) is with us as we witness the battle.

The imaginative activity of the perceiving subject works with Welles's
rich dialectical structure of inside and outside in establishing the spatial
field. Court and tavern combine to form a texture of retreat, the importance
of which is clear only when the dangers of the outside are apparent in the
Shrewsbury sequence. The dynamic of stasis and movement in court and
tavern respectively finds a counterpart in the outside in the horrifying
stillness of death after the moving charge of battle. Court intrigue and
politics find expression in the battle of the outside, take form in the outside,
just as Lear's madness finds articulation in an external space of action. The
playful games of tavern are a foreshadowing of the "game" of war, a juxtaposition that reveals another side of human playing and sheds light on the
more serious convictions of a Hotspur, a Henry, or a Worcester—all caught
up in a mad game of power. Falstaff's lies "inside" are a foreshadowing of
those told by Worcester; the Knight's recruitment of and bribe-taking from
his company of pathetic soldiers are acts that must unfold in relationship to
their terrifying death. In these ways the inside informs the outside and
serves to point out significant relationships between seemingly antithetical
worlds. The violence of the outside penetrates the inside because it opens it
to a new kind of scrutiny. Outside battle and inside policies; outside bloodshed and inside intrigue; outside suffering and inside definitions of honor,
loyalty, and obedience; outside action and inside talk—all work to create,
within the spatial field, the tensions on which the drama operates.

Finally, in a manner similar to that of Kozintsev's *King Lear*, outside and
inside, in all their complex relationships, reflect important tensions in the
central figure himself. One of Welles's great achievements in the film is his
realization of the ambiguities of the Falstaff figure. As Dover Wilson points
out, "Falstaff is a bundle of contradictions." Building on this claim, Empson
adds that "the whole Falstaff story needs to be looked at in terms of Dramatic Ambiguity."[10] In Shakespeare's text, Falstaff is a figure in whom
many disparate attributes meet; his physical size gives form to the scope of
the contradictions he can accommodate. Out of the mouth of this man
comes a request to Hal (his first request) in Act 1, scene 2 of *1 Henry IV* to

make thieves "men of good government." At one moment he speaks of giving his soul over to God, and, at the next, he plans the Gadshill robbery, moving quickly from "praying" to "purse-taking." In the Knight's estimation, Hal's royalty depends on his willingness to stand for ten shillings so that "the true prince may . . . prove a false thief." He continues to equate thievery and honesty in his memorable remark, "A plague upon it when thieves cannot be true one to another" (2.2.26). In the second part of *Henry IV*, he refers to himself as both "slender" and "great." As the man who conceives of the world in terms of discrepancies, he asserts that he "will turn diseases to commodity" (1.2.247–48). The myriad sides of Falstaff find expression in his signature on his letter to Hal. He is many men at once: "Jack Falstaff with familiars, John with my brothers and sisters, and Sir John with all Europe" (2.2.125–27). For Falstaff, a firm negative must be the answer to his most basic question: "Is not the truth the truth?" (2.4.227). Such equations are not possible in the Knight's ontology.

Shakespeare also evidences the ambiguities of Falstaff's character in the way the Knight conceives of himself as young and old simultaneously. He includes himself in the company of "us youth" in both parts of *Henry IV* (Part 1, 2.2.85; Part 2, 1.2.172–73), and yet the playwright juxtaposes the Knight's playful, childlike spirit with his nagging awareness of age and mortality. Dramatically, Falstaff is both youth and age; he expresses this most clearly in his words to the Chief Justice: "I was born about three of the clock in the afternoon, with a white head and something of a round belly" (1.2.185–87). Moreover, in the scene with Doll Tearsheet, we witness a figure not only young and old but both amorous and impotent as well.

Other characters comment on Falstaff in contrasting terms. For Hal he is both humorous and pathetic: "were't not for laughing, I should pity him" (2.2.99). His reputation is ambiguous; all know him as a thief, and yet the Chief Justice recognizes him in Part 2 as having "done good service at Shrewsbury" (1.2.61). In the words of the Chief Justice, Falstaff has the ability of "wrenching the true cause the false way" (2.1.107–08). Moreover, on many occasions, Falstaff shows two faces simultaneously. This is most evident at such moments as his recovery from the Gadshill lie (Part 1, 2.4); in his rejoinder that his love for Hal is "worth a million" (Part 1, 3.3); or in his show of friendship towards Hal and Poins after they have listened, in disguise, to Falstaff's vituperations about them (Part 2, 2.4). For almost anything one can say about the Knight, the opposite is usually true as well. His actions show him to be both lovable and repugnant, cowardly and stouthearted, a liar and a man honest about his failures. His language is both witty and ridiculous, intelligent and empty. In short, Falstaff appears to both the audience and the characters who interact with him as a wholly ambiguous figure. I stress the point in light of the fact that Welles centers his film, and guides the spectator through the world of the *Henry IV* plays, on the basis of a "bundle of contradictions."

As a film that focuses on the journey of Falstaff, *Chimes at Midnight* offers in the inside and outside a series of contradictions that work to define the spatial field in a manner similar to the way we work with many parts to characterize Falstaff himself. Welles's spatial field, along with Shakespeare's own design of the Falstaff character, operates in a dynamic of relationships between retreat and violence, stasis and movement, praying and purse-taking, governing and thieving, truths and falsehoods, diseases and commodities, age and youth, mortality and playfulness, pathos and humor. In addition, just as the experience of Kozintsev's Lear from inside to outside reflects the psychological journey of the King, so, too, do spatial configurations parallel the experience of Welles's Falstaff. As we move in the film from the good fun of the tavern to the horrors of Shrewsbury, we simultaneously trace the gradual psychological degeneration of the central figure. Falstaff's deterioration throughout the course of the film is reflected in the movement from inside to outside, and the spatial field of the film tells us as much about the experience of the fat Knight as it does about the larger historical moment.

III

The dialectic of inside and outside also articulates the social character of film, another function of the spatial field of cinema. Filmic *mise-en-scène* is a field for performance made up of a series of relationships that formulate a social discourse as much as an aesthetic one. I pointed out in the discussion above how the outside of Lear's kingdom was informed by the inside world of politics, and that the suffering of humanity beyond the court expresses an important aspect of the social character of the inside. A similar dynamic was evident in *Chimes at Midnight*. To put it another way, inside and outside work to defamiliarize the political structure of each other, and by so doing offer the viewer material from a specifically social and historical point of view. Indeed, inside and outside function in precisely the way Brecht wanted all elements to function; they work together to "alienate" the audience, to make the action "distant," to "make strange" the familiar setting, to inspire the spectator to ask new questions about the socioeconomic context of the drama. In what follows, I consider the dialectic of inside and outside as part of that which realizes, dramatically, a sense of the historical process operative in Kozintsev's *Hamlet* in order to understand not only the aesthetic and psychological implications of spatial dynamics, but the social ones as well.

In 1954, when Kozintsev directed a stage production of *Hamlet* at the Pushkin Academic Theatre of Drama in Leningrad (a production that made its mark on Soviet theater history) he craved what only the cinema could give to the play:

> I am a movie director, and often, in the middle of the beautiful theater which Rossi had designed, I missed the northern winds and the broad spaces of the sea. . . . The opportunity of looking into the very depth of Hamlet's soul was not enough for me. I desperately needed a movie camera.[11]

I would suggest that Kozintsev was "desperate" for the multiple perspectives that cinema offers and the range of spaces that it can cover because film gives the Shakespeare play a unique context that, among many things, allows for a specific exploration of the social character of the drama within the historical process. As a student of Eisenstein, Kozintsev uses the film medium to aestheticize Marx's concept of dialectical materialism. While Eisenstein relied mostly on montage as the aesthetic form that best mimics Marxian historical process, Kozintsev goes a step further in his Shakespeare films by contextualizing the activities of the political regime within a larger whole. His filmmaking strategy exploits the medium to show the operation of historical forces, and he achieves this, most convincingly, in his use of cinematic space and in the specific juxtaposition of inside and outside. The inside space of human activity and politics unfolds in a relationship to an outside world of "northern winds and the broad spaces of the sea." The relationship is one of outside movement and inside stasis, of natural freedom and synthetic contrivance, of a world that accommodates past, present, and future and one built to isolate time in the seemingly unchangeable moment of a fascistic regime.

In a paper delivered to the 1971 World Shakespeare Congress, Kozintsev remarked that "the process of tracing the spiritual life of Shakespeare's plays cannot be separated from the tracing of the historical process."[12] He constructs his *Hamlet* precisely along these two lines of action; his film is concerned with the working dynamics of political power and the effect of power on the human spirit. What *Hamlet* announces is a specific conflict in history, nourished both by the activities of Claudius and his court and by the inevitable forces that collide with those activities. It thus shows the disease of the time, a monstrous moment in history, as well as the larger process of which that moment is only a part. To express the issue in terms of *mise-en-scène*, Kozintsev explores the tragedy of a man who is destroyed by the collision of inside and outside.

The prologue to the film immediately establishes the spatial field as a dynamic of the closed and the open, of stillness and movement. Kozintsev offers in his first shot an aerial view of the sea surrounding the enormous cliffs of Elsinore. We do not see the castle but, instead, its image projected on the water. The outside and inside appear initially as sea and shadow, as a two-dimensional, static shape superimposed on a three-dimensional moving surface. (The black shadow of Claudius's deeds casts itself on a world that extends beyond the microcosm of Elsinore.) The director fills the outside world of this film with images of movement, of sea and sky, of clouds continually forming new configurations, of birds circling in the sky, of wind, and of waves crashing on the shore. It is a world that gives visual and aesthetic form to that part of the historical process the human agent plays no part in, moving at its own pace (but always moving), vast, enigmatic, independent of synthetic elements, a world compiled of only natural phenomena. In juxtaposition to this outside is the static, political trap that is Elsinore, made explicit in the raising of the castle drawbridge (an image of isolation) and the simultaneous lowering of a huge iron portcullis after Hamlet enters on his return to Denmark from Wittenberg. The portcullis slams shut on the inside world as if to quarantine the disease of Claudius's tyranny.

From the opening visuals alone, the character of Elsinore emerges in contrast with the surrounding outside world. The shadow it projects is a sign of stasis—not in and of itself, but in relationship to the motion of the sea that serves as its backdrop. As we move inside the castle, Kozintsev echoes in his *mise-en-scène* comparable relations between movement and stillness as he builds the spatial field of Claudius's regime. The film echoes the two-dimensional nature of Elsinore's shadow outside with a series of tapestries that cover the walls of the inside. They depict scenes of slightly distorted and lifeless figures in acts of hunting and war, all set against a flowery background that suggests an incongruous context of benevolence. The figures of the tapestries find their walking counterparts throughout the court in the form of Claudius's politic worms milling about. Moreover, the artificial images of hunting and war find a parallel in the violence of the King's deeds, also couched in benevolent seeming. Always with an eye on the spiritual rupture that results from political oppression, Kozintsev will offer moments in which a character seems to become one with the lifeless figures of the tapestries. When Laertes gives Ophelia advice on her relationship with Hamlet, he speaks to her as she leans against the wall. The stifling nature of Laertes's counsel (a political act of oppression that has as much to do with the exploitation of Ophelia as a woman as it does with internal politics of court and family) strips Ophelia of her dignity, and she seems, visually, to merge with the artificial decor of Elsinore. Kozintsev uses the tapestries for a similar purpose in the closet scene: "The development of her [Gertrude's] character is made clear by a shot in her bedroom as Hamlet enters—as she stands near the hangings, her dark silhouette makes it seem that she has been cut loose and lifted out of their two-dimensional world."[13]

Other details of *mise-en-scène* suggesting the static character of the inside include the director's myriad statues, dress-forms (for Gertrude's wardrobe), shadows, the tense and rigid poses of Claudius's courtiers, and the hard and immobile textures of stone and iron prominent inside. But the movement of the outside finds its way indoors through the hero himself. Hamlet is a moving fire within the stone and iron of Elsinore and, as such, he threatens the political stability of Claudius's regime. From the very first image of him galloping on a white horse, moving against an ominous sky of thick black clouds as he journeys towards the Elsinore fortress, Kozintsev's Hamlet is a figure always on the move, restless, and associated with fire, sea, and sky (elements that repeatedly appear in shots of him). He is the only major figure to appear on several occasions outside of the fortress walls; he even goes outside to die at the end of the film. Hamlet carries with him the outside forces of movement and is their ambassador within Claudius's tyrannical order. In addition, Hamlet has left Wittenberg for Elsinore, which, in Kozintsev's scheme, was a departure from the new humanism of the outside to the feudal barbarism of the inside, a departure to a political reality that makes the advance of truth and knowledge a fruitless and inconsequential quest in the face of tyranny. "In many eras, the finest men knew despair; lofty dreams proved so obviously futile. The time came when the heavy cannon of the 'Elsinore' of the times dispersed the ideas of the 'Wittenberg.' "[14]

Ironically, the most "concrete" force of the outside world is the Ghost, and Kozintsev arranges his text to suggest that its appearance brings on a specifically political conflict in the tragedy. The Ghost, as a force of the "outside," is a threat to the political regime of the inside and gives the lie to Elsinore's seeming impenetrability. Kozintsev cuts Act 1, scene 1 from the film (delaying the Ghost's first entrance to 1.4) to demonstrate that human action is the source of Denmark's depravity. In addition, without this scene, the Ghost no longer appears in a religious context. Gone is the moment when the Ghost is "offended" and "stalks away" after Horatio charges it "by heaven" to speak. Gone are Horatio's references to the Ghost as a possible spirit of evil when he calls it "illusion" or when he determines to "cross it" though it may "blast" him. Gone, too, is the Ghost's departure at the sound of the cock, which Marcellus explicitly throws into a Christian context by associating it with "so hallowed and so gracious" a time (Advent) when "no spirit dare stir abroad."[15] Indeed, Kozintsev removes Shakespeare's many clues in the play that tell us that the Ghost needs to be understood, initially at least, within a Christian framework. Instead, he shows first an inside world of stasis: shadows and tapestries, a huge fortress of iron and stone, soldiers in armor, a portcullis, and a moat that ensures physical isolation. What Shakespeare emphasizes in his opening scene through the mysterious appearance of the Ghost and its possible malevolence ("a goblin damn'd") in a religious context, Kozintsev replaces with a prologue that defines the "strange eruption" in social and political terms; the onus is not on some supernatural force but on politics itself. Whatever may be "rotten in the state of Denmark" is caused by the human beings who inhabit it. Only then is the stage set for the entrance of the Ghost.

The Ghost, when it does appear, gives a temporal definition to the outside, a definition that enhances the disparity between Elsinore and what is beyond. In Kozintsev's film, the Ghost is an essential ingredient in the dialectical movement of history, and, as a presence initially "of the outside," it finds its way inside by challenging Claudius's attempt to freeze history and time in its own tyrannical order:

> The dead king comes into the tragedy from the stern sagas of the past. He is foreign to the new epoch and its morals. He seems to be a reminder of what had once been knightly valor, honorable relations among sovereigns, the sanctity of throne and family.[16]

Kozintsev's Ghost is a force of the past materializing in the present, but it is also a sign of the future, a "herald of natural disasters . . . [and] a typical figure that gives forewarning of events."[17] Kozintsev replaces Shakespeare's specifically religious context with a historical context that makes the Ghost a complex symbol of time, a force of the historical process (in which the scope of time is accommodated) that collides with a king who tries to tyrannize them, to isolate his regime in history as he isolates his fortress from all that surrounds it. The historical process thus takes on spatial dimensions. Instead of Horatio's specifically Christian references, we have Kozintsev's close-up—just before the Ghost enters—on a clock striking the hour of midnight, the moment when beginning and end are one, "the season / Wherein the spirit held his wont to walk." Figures of a bishop, a king, a queen, a knight, and Death itself circle below the face of the clock and create a specific framework for the Ghost's presence: the icon of Death follows icons of warfare, royalty, and religion. The circle continues. History follows its pattern. Time moves. And that movement will find its way into Claudius's static world.

The spatial conflict of inside and outside, then, is emblematic of the conflicts of the historical process, of movement and stasis, of past and future with the present, of a "new rationalism" in Wittenberg and feudal oppression of Elsinore. A primary characteristic of the inside is its flagrant attempt to cut itself off, to isolate itself from the myriad forces of the outside—an attempt at exclusiveness that defines Claudius's despotism. "A modern

Elsinore would have no objection to closing the barbed wire of concentration camps around humanity like a crown of thorns."[18] Despite Claudius's attempts to keep the outside world at a distance, despite his drawbridge and portcullis, despite his attempts to mold Hamlet according to the political dictates of his interior regime, the outside works its way inside and the collision results. Just as Lear's and Falstaff's experiences found articulation in the inside–outside dynamic of Kozintsev's and Welles's respective *mise-en-scènes*, so, too, does Hamlet's. Hamlet is of both worlds: Wittenberg and Elsinore, movement (thinking) and stasis (his delay), the Ghost and Gertrude. Ultimately, though, Hamlet sets events in motion. He challenges the static interior with movement; he brings with him the uncompromising spirit of the outside, a spirit that takes the form of thought, of questioning, of political challenge. Claudius is afraid. "Hamlet . . . thinks. There is nothing more dangerous."[19]

We are aware of Hamlet as the moving force from the beginning, even before he confronts the Ghost. In the assembly (1.2), all of the nobles sit obediently in Claudius's court while he looks after his business. Hamlet sits alone at the far end of the table. When Claudius finally turns to the Prince ("How fares my cousin Hamlet, and my son?"), the camera pans the long table of nobles to reach an empty chair. Hamlet is gone. What was stock still must now move; the court, led by Gertrude, chases Hamlet. The scene then continues "on the move" by way of a traveling camera. Gertrude catches up with Hamlet and walks with him while asking him to cast his "nighted color off." (They briefly stop on a landing of a staircase for Gertrude to primp in front of a mirror, an activity that takes place, significantly, during Hamlet's "seems, madam" speech.) Claudius finally joins them, announces Hamlet as the figure "Most immediate to our throne," asks him not to return to Wittenberg, and comments that "This gentle and unforc'd accord of Hamlet / Sits smiling to my heart. . . ." Several courtiers and servants—who have been following Claudius like a flock of sheep—shout in spineless unison: "Long live the king." Hamlet disturbs the stillness of the inside; he sets the court in motion.

In one sense, Hamlet's dilemma in the "to be or not to be" soliloquy highlights the tension between inside and outside, between movement and stasis. Interestingly, Kozintsev employs his medium to shoot the speech "on the move" in the outside world; Hamlet walks along the seashore, negotiating its rocks and dodging the waves washing on the beach as he contemplates the question of action. We move with the camera as it moves with the hero in a compelling blend of contemplation and motion: "The whole point was to link the rhythm of the cine camera's movements with the main character's thoughts."[20] The dialectics of inside and outside inform Hamlet's thoughts and frame his question: must one move and take action or must one wait and be still? Hamlet can capitulate like the politic worms of Elsinore or find the strength to stand against tyranny.

But Hamlet's burden—to remedy a time out of joint, to stir the forces of history, to carry the conflict of inside and outside—is too great for any one individual to bear. The tragedy in Kozintsev's film is evident in the imposing of the responsibility on one man, despite his heroic stature, to reconcile forces at play in history; it is an imposition that ultimately leads to spiritual, psychological, and physical collapse. "This tragedy portrays a man who does not find himself between life and death, but between one era and another."[21] Hamlet brings with him a certain dignity of the outside, a sense of the possibility of progress to another era; he will not be played upon, he will not acquiesce. His tragic stature is his ability to challenge, to move in stasis, to think, to be ready. And in that process he learns:

> the Wittenberg student discovers a new and unfamiliar world, types of people which had been unknown to him, social relations and moral norms which he had not learned. Here, in Elsinore, the life of the times moves before Hamlet's eyes. . . . Hamlet discovers the world in which he lives. He discovers the souls of his mother and his beloved, the consciences of his friends, the moral philosophy of his courtiers. He discovers himself. And then he dies, because one cannot live on, having learned all that he has.[22]

History moves through conflict. History is a dialectic of forces colliding to move humanity forward. Hamlet, as a figure who has the capacity to carry the conflicts of his time, moves forward in discovery and learning and finds nourishment from the same dialectical clash as that which advances humanity in historical time. But while history can survive the radical development stemming from conflict, the individual human being—"having learned all that he has"—cannot. He can only be an agent in the process; he can only be "ready." Hamlet kills Claudius. Fortinbras enters from the outside. The final image of the film repeats the very first one with a picture of Elsinore's shadow projected on the sea. We are thus left with a question: has there in fact been any advance or will the advent of Fortinbras portend a new era of terror?

With elements interacting on many levels, the *mise-en-scène* is, of necessity, a structuring by the viewer of the materials that the film offers. Obviously, this act of structuring follows certain guidelines created by the artists involved (Shakespeare, Kozintsev, Welles, the actors, designers, etc.), but the elements they create are, finally, linked in the imagination of the perceiving subject. Inside and outside work to form the visual poetry of the film in the act of viewing, and the relationship between the spatial field and the journey of the central figures is a dynamic that finds nourishment only from that act. "This notion of *mise-en-scène* operates a radical transformation, moving from the finalized exterior object to the structuring effort of the perceiving subject. It has become a structural principle of organization which generates and creates the performance from projects/propositions of the stage [and screen] and responses/choices of the audience."[23] The spatial field is thus built on a series of relationships in the virtual dimension of the film. The imagination shapes the raw material of the film, raw material that the filmmaker offers as an open structure to be organized in the viewing process.

Curiously, in that viewing process, the spectators are both inside and outside the experience of the film. They observe the events on the screen, participate in the illusion, "move" with the images, observe many points of view, and yet sit in their fixed positions throughout the entire experience. The spectator observing relationships of inside and outside, of movement and stasis, is observing a mirror of his or her own physical and imaginative activities in the cinema (not to mention the fact that the medium itself is built on the "movement" of a series of "still" pictures). Stephen Heath clarifies the point when he writes that "What moves in the film, finally, is the spectator, immobile in front of the screen. Film is the regulation of that movement, the individual as subject held in a shifting and placing of desire, energy, contradiction, in a perpetual retotalization of the imagery (the set scene of image and subject). This is the investment of film in narrativization; and crucially for a coherent space, the unity of place for vision."[24] As the films unfold in a relationship of stasis and movement and of inside and outside, they comment on and reflect the very process of viewing and the dynamic of the physical and imaginative processes of film.

Houseless Heads: The Storm of *King Lear* in the Films of Peter Brook and Grigory Kozintsev

Historically, the storm scenes of *King Lear* have stimulated a debate about the relative virtues of reading Shakespeare's play and seeing it performed. The rich experience of private contemplation of the playwright's poetry, limited only by the space of imaginative musing, collides with the relatively crude realization of that poetry in performance. Charles Lamb is notorious for his statement that "the Lear of Shakespeare cannot be acted," and for his assertion that "the contemptible machinery by which they mimic the storm which he goes out in, is not more inadequate to represent the horrors of the real elements, than any actor can be to represent Lear."[1] A. C. Bradley, too, denounces the storm as a poor theatrical device and defines the problem as a conflict between "imagination and sense," between the boundless freedom of the mind and the hard limitations of the stage: the storm scenes are "poetry, and such poetry cannot be transferred to the space behind the footlights, but has its being only in imagination."[2] Lamb and Bradley are significant here, not for their audacious call to end all productions of Shakespeare's tragedy (though one can understand their objections, given the kinds of productions they might have seen—with a growing emphasis in the nineteenth century on pictorial realism), but for the tension they perceive between imaginative and real experience. Today, most of us would probably respond to their challenge by emphasizing how performance activates the imagination in a manner different from that of the reading process. We would probably agree with Maynard Mack when he writes that the best production of *King Lear* is one that works with Shakespeare's play to "produce a storm in the audience's imagination," or with Marvin Rosenberg, when he argues that the storm scenes in the theater are inadequate only "if poorly done."[3] If sense and imagination collide for the spectator of the storm scenes, it is the fault not of the poetry but of the production itself; if the director employs pyrotechnics that distract the audience, if he or she overwhelms the spectator with details of wind and rain, if the scenes focus on making a "realistic" storm in the theater rather than on the questions of human experience asked within the context of Shakespeare's tragedy, then bad theater is the root of the problem, and not poetry on the stage.

Another way to look at the issue is to ask how the director, on stage or film, constructs a context for the events of the storm and how the conceptual focus of the production informs that context. The relationships built by the director between Lear and his environment, between a tragic figure and a world in chaos (whether that chaos be a representation of the internal struggles of the character, or the physical world in upheaval, or both), are the central factors activating the imaginative participation of the audience. The stage director who understands the creative potential of the medium will use the theater, not to create a realistic storm, but to offer a context for the turbulent confrontation of a man and the elements.

A central issue in translating Shakespeare to the screen has to do with the extent of the filmmaker's exploitation of the technical resources of the medium to present the plays according to the conventions of "photographic realism." Implicit in the statements of Bradley and Lamb is a criticism of the theater's feeble attempt to realize the poetry of the storm, a realization that they perceive only as mimetic reductionism, no matter how elaborate the attempt at pictorial realism. The boards will always limit. But the filmmaker has at his or her disposal the technical resources to imitate the details of the storm in a highly developed manner and thus to satisfy, potentially, frustrations like those expressed by these two critics. The film medium has the capacity to create a real storm, with real wind, rain, and thunder. The challenge of cinema, however, is no different, fundamentally, from the challenge of stage: the context of action, the relationships among key elements, is the crucial factor in the creative process, not the degree of realism achieved. Indeed, with an overly realistic filming of the storm without the framework that inspires the imaginative activity of the spectator, the objections of Lamb and Bradley hold as much for the screen as they would for the stage. We would have a crude reduction of dramatic poetry in either case.

A study of film treatments of the storm scenes, therefore, is suggestive because one can address the issue of how the filmmaker employs the medium to exploit its unique resources; one would hope, without quelling the imaginative resources of the viewer. To this end, the treatments of the storm by Brook and Kozintsev, though fundamentally different in conceptual focus and in the specifics of how they use the medium, both give evidence of the potential that film offers this moment in Shakespeare's tragedy. Both directors achieve their ends by capitalizing on the spatial possibilities of the cinema to create the context. Whereas Kozintsev emphasizes the change in the material and physical world in which we perceive the hero, Brook presents the events within the psychological space of the troubled King. In the one film we recognize the new spatial contexts that define the human being in relationship to his physical environment (and how spiritual change stems from that confrontation); in the other we see the new relationships that define the human being in the context of a tortured psyche. To express the issue as a paraphrase of Marx and Engels, Kozintsev suggests that the environment fosters growth and determines the new consciousness of the King, whereas Brook shows how the King's troubled mind determines the world. Most pertinent to my concern, however, is the fact that a study of film treatments of Shakespeare, no matter what the conceptual focus, is, on one level at least, a study of the new relationships that cinematic space creates.

Kozintsev writes that "in order to create poetry on screen you must first of all discover the prose in it."[4] In other words, the first relationship concerning this director is the one between elements that ground the spectator in an everyday context and those that allow for the growth and movement that arise within that context. The specific framework of the storm scenes in the Soviet film is the new space that Lear enters to discover the landscape of his kingdom, a space the director makes vivid through a fundamental change in Lear's relationship to the world around him. The emphasis for Kozintsev is not so much on the more traditional notion of the storm as representing both psychic breakdown for Lear and physical chaos of the natural world,[5] but on the dialogue unfolding between a human being and the elements.

> The usual interpretation of this scene as representing the chaos of the hero's spiritual world through the boisterousness of the elements seems to me to be literary and static. Here the most important things are movement and conflict. The main concern is not chaos (madness), but the fight against chaos; what is the reason for breaking all links, for the inhumanness of it all? Where are the roots of evil hidden? Lear converses with the elements straight-forwardly, as an equal; he demands their sympathy and gets angry with them for being heartless.[6]

Kozintsev builds to the poetry of the storm by first establishing the "prose." Only then does he put the hero in a new spatial field of poetic action. The sequence begins with the everyday action of attendants' closing the doors of Gloucester's castle behind Lear and the Fool. The servants, in "prosaic" activity, pull down clothes that have been drying on a line. One sees all of this in close-up; the human figure still looms large on the screen. Kozintsev offers additional signs of the storm with shots of a darkening sky, of wolves, of bears and boars roaming the woods seeking shelter, and of a herd of wild horses galloping on an open plain. "And then the prose changes to poetry. Man no longer walks through the rooms and lands of an estate, but through the threatening lands of tragedy. The proportions have suddenly altered, man is insignificantly small, powerless, defenseless, and the expanse is enormous and hostile."[7] In a distant aerial shot, Lear is a tiny shadow of a man alone on a huge, desolate field. In contrast to the close-up shots of the King in Gloucester's castle (evident in the sequence preceding the storm), the first shots of the storm scene reveal the stature of the unaccommodated human creature.

Kozintsev contextualizes the action through spatial relationships. In the space of the close-up before the storm—in the civilized world of his throne room, Goneril's castle, a small carriage he rides in, or Gloucester's castle—Lear looms large on the screen. On the heath he is tiny, insignificant, alone. For the spectator, the imaginative experience of the storm in Kozintsev's *King Lear* is an experience of the action as a dynamic within the spatial field. In addition, the filmmaker does not exploit the resources of his medium to portray a man overwhelmed by the forces of nature; we do not see Lear mercilessly destroyed by rain and wind—that would ruin the imaginative participation of the spectator. Kozintsev offers only enough to establish the prosaic foundation out of which the poetry can grow. The prose of wind and rain and the words of a man turn to poetry through interaction, through a dynamic of cries both of man and of nature, through a "conversation" between a suffering, enraged king and the angry gods.

The spectator assumes, temporarily, the gods' point of view as Lear cries out, "Blow, winds, and crack your cheeks! Rage, blow!" Lear challenges the storm and it answers back with all its thunder. The aerial shot enhances the feeling of dialogue between Lear and the elements because the camera forces us to share not the King's more familiar perspective but the unfamiliar and hence challenging perspective of his interlocutor—the skies. Finally, with the entrance of Kent, the camera moves down to the ground. Removed from our privileged, godlike perspective, we are with Lear again as he assumes normal size and proportion on his line, "I am a man / More sinn'd against than sinning." He holds the Fool to his breast: "Come on, my boy? How dost, my boy? Art cold?" The three men then walk off to the hovel.

Kozintsev contrasts the barren and empty landscape of the heath with a hovel in which a multitude of naked wretches cram together seeking protection from the storm. In Kozintsev's film, Lear finds not only one suffering figure in the face of Poor Tom, but an overwhelming mass of "poor, bare, fork'd animal[s]." In his disguise, Edgar shares the space, it seems, with all houseless poverty. In fact, he decides to assume his disguise after seeing the wandering peasants of the outside world; their wandering and suffering inspire the exiled Edgar to become one with them, to feel what they feel, and to seek shelter from the elements in the space of their misery. Indeed, the "space of tragedy" in Kozintsev's film is the space of affliction, a space that the camera reveals confined and crammed with humanity, a space that the spectator enters with the aid of the camera so that we share intimately in the pain all around. It is the space, moreover, that provides the context for Lear's profoundest discoveries.

In the film, as opposed to the original text, Lear's actual perception of "loop'd and window'd raggedness" is the impetus for his prayer for the houseless beggars and for his call to the rulers and privileged ones of the world to heal themselves with the "physic" of suffering. Interestingly, Shakespeare purposely delays the entrance of Poor Tom so that the King's prayer occurs before he confronts "the thing itself." In the play, Lear's suffering triggers an awareness of what it must be like for the "poor naked wretches" of the world to endure storms such as these; his breakthrough is an imaginative one resulting, not from a direct confrontation with them, but from his own experience—he calls out to his fellow human beings with the words "wheresoe'er you are." Only after this preparation by the elements does Lear find Poor Tom. Kozintsev follows Shakespeare's delay of Lear's confronting Poor Tom (he hides under a pile of straw when Lear enters the hovel), but creates a specific context for the "prayer." Surrounded by the refugees of the hovel, Lear asks:

> How shall your houseless heads and unfed sides,
> Your loop'd and window'd raggedness, defend you
> From seasons such as these? O, I have ta'en
> Too little care of this! Take physic, pomp;
> Expose thyself to feel what wretches feel,
> That thou mayst shake the superflux to them,
> And show the heavens more just.
>
> (3.4.30–36)

The "poetry" of Lear's spiritual opening has roots in the "prose" of real human beings in a space of misery. Though Kozintsev may lose the staggering power of the imaginative discovery of Shakespeare's hero, his shift of context is important for many reasons. First, with his camera, Kozintsev takes us inside and throws us into the confined space of the hovel, putting us in the immediate world of Lear's experience; we thus share the space with Lear and witness, from that intimate perspective, the pain that fills that space. Second, Kozintsev's Marxian orientation leads him to present this speech (the words of a king who calls for distributive justice) within the context of real human beings with real material needs. Lear's discovery is not an abstract, philosophical, or theological one (another reason for Kozintsev's having the King say his "prayer" directly to other human beings and not to an invisible god), and Kozintsev takes pains to make the King's learning process as visceral as possible within the performance. Finally, Kozintsev contrasts this moment with key moments that precede it. Masses of humanity have always surrounded the King; only now does he see them in terms, not of his own gratification, but of their needs as people, needs the King must tend to. The spectator recalls the crowded court of the love contest as he or she witnesses the space of the hovel. Towards the end of the scene, the beggars of the hovel become the jurors for Lear's mock trial in a sequence that ends with the King's turning to one of them, holding him, and asking, "Is there any cause in nature that make these hard hearts?" Kozintsev creates a context for one of the most fundamental questions of the tragedy by showing the King, in close-up, searching for answers in humanity's lowest creatures.

The context of Kozintsev's storm, therefore, is evident in the vast space that renders the human figure tiny, in the balance of languages of man and the elements as they converse, and in the close and intimate space of the crammed hovel. We observe Lear's experience in relationship to a world and to events external to him. His growth, his learning, and his expanded consciousness are direct products of the environment he confronts. By contrast, the emphasis of Brook's technique is not so much on rendering Lear's experience within a specific external framework as it is on using the resources of the cinema to give form to the King's disorientation and madness. Kozintsev creates a relationship of activities between character and milieu; Brook creates one between character and tortured mind. In one film the emphasis is on the parallel course of the human spirit and history; in the other, it is on the journey of the deranged individual.

Our first clue in Brook's film that what we are to witness is the subjective journey of the hero is apparent in the departure from Gloucester's castle. Chaos ensues as the camera whirls in search of Lear. When the camera does find him, he appears lost, searching for his carriage, his horse, his serving men, his fool, his sanity. The sounds and visuals of blowing wind provide the filter through which we observe the events, a filter of Lear's own incipient madness. Brook's storm is the storm Lear perceives; the treatment of the departure from Gloucester's castle, in all its disorientation, follows immediately upon the moment when Lear whispers to his Fool, in the private space of the close-up, "I shall go mad." That announcement is the prelude for all that follows. This is a storm of raging wind and rain drowning the world, of "oak-cleaving thunderbolts," of "germains" (floating in pools of water) that make ungrateful men, of animals unable to survive, of naked wretches with no shelter. But Brook does not simply repeat in the visual language of the cinema the words of the playwright; he creates the state of mind out of which those words find expression. Brook's visuals act as a filter of subjectivity; they suggest *how* the character sees before they describe *what* he sees.

Peter Brook's *King Lear*. "Who is it that can tell me who I am?"

Peter Brook's *King Lear*. Lear and his knights departing Goneril's castle.

Peter Brook's *King Lear.* Paul Scofield as Lear and Jack MacGowran as The Fool.

Following his departure from Gloucester's castle, Lear drives his carriage wildly through the storm, whipping his horses to gallop at full speed. His movement is his madness. The storm on film affects, literally, our perception of the world as the torrents of rain blur the shots of the King on his speeding carriage. A wheel on the carriage breaks; he mounts one of the horses and rides it bareback; the horse falls, unable to bide the storm; he leaves the horse and walks up the heath, rain pelting him; the Fool struggles to follow the King; thunder sounds shatteringly and all goes black. Then, in the black nothingness sounds the voice of Lear: "Blow, winds, and crack your cheeks!" We hear another clap of thunder and the black frame turns to white, unveiling the face of Lear in close-up. He is lying on the ground and we perceive him in a watery out-of-focus image as he continues his speech. Lear gets up on his feet and cries out to the sky; his arms stretched toward heaven are visible to us in a low-angle shot that emphasizes the impossibility of his reach. Black frames return momentarily; the wind and rain in both the aural and visual fields are unending, merciless forces. The Fool hides his head in an animal hole. When Kent finds the King, Brook presents Lear's speech on exposing the vicious of the world—"Let the great gods, / That keep this dreadful pudder o'er our heads, / Find out their enemies now"—in a sequence of alternating shots of Lear in profile on the left side of the frame

Peter Brook's *King Lear*. Lear, in the storm, cries out to the heavens.

and shots of the opposite profile on the right. The effect is to create the illusion of Lear interrogating himself (as he "faces" himself in dialogue), of a king who examines himself from both sides, of a man whom we see exposed from more than one angle. Lear observes himself as "the wretch that trembles," while the alternating shots realize the very process of self-scrutiny. Kent then leads King Lear towards a hovel.

At this moment, on "But where the greater malady is fix'd, / The lesser is scarce felt," Brook takes a full-frame close-up of Lear's face, hammered by rain and wind. The shot pounds Lear in close scrutiny, matched in power by the rain hitting him and dripping from his face like tears that could come only from a giant's sorrow. We see just a part of his face in close-up on "The tempest in my mind / Doth from my senses take all feeling else / Save what beats there." Brook uses the partial shot of the face in close-up throughout the storm scenes as a primary image of fragmentation and the loss of self. Suddenly, Brook cuts to a long shot of Lear in a void of blackness, appearing tiny in context with only vague details of his face and white beard giving definition to his being. He cries of "filial ingratitude." A climax of the journey then occurs with his prayer for the "poor naked wretches," this time through a sudden break in all the storm sounds (ineluctable to this point) and the isolation, in the aural field, of Lear's voice in prayer. As we listen to the speech in voice-over (the only sound in the startling silence) we see an overwhelming vision of dead animals in a rushing stream, a monument to this unforgiving storm. By giving the aural and visual fields sepa-

rate spaces, Brook invites us to see the world with the King. We listen only
to the internal voice, a voice that teaches caring for humanity and shaking
the superflux to the poor.

Lear meets Poor Tom and asks him, "What hast thou been?" Tom answers in a medium close-up shot (standing in near-naked, Christlike agony),
but Brook punctuates every phrase he utters with a clap of thunder and an
immediate cut to a full close-up of this tortured figure.

> A serving man proud in heart and mind; [thunderbolt and close-up] that curled
> my hair; [thunderbolt and close-up] wore gloves in my cap; [thunderbolt and
> close-up] serv'd the lust of my mistress' heart, and did the act of darkness with
> her; [pattern repeats to the end of the speech] (3.4.84–87).

The director thus makes the image of Tom both real and a phantom of
imagination, an artificial construct created by myriad participating forces:
Lear, Edgar, Shakespeare, Brook, the storm, and the viewing audience.
That Tom is, on one level, a construct of Shakespeare and Brook is obvious.
Subtler is the fact that Tom is a construct of Lear (we perceive him through
the filter of the King's consciousness in close and medium shots), a construct
of Edgar as an act of survival (Tom, on one level, is Edgar's invention), and a
construct of the storm in the sense that his suffering is a product of the
"extremity of the skies." Ultimately, however, he is a construct of the spectator working with all the forces that constitute Tom's being. Brook penetrates Lear's face once again at the beginning of the speech in which the
King asks, "Is man no more than this?" But the shot is slightly out of focus
and we cannot see the details of the King's eyes; we recognize only the
deepset hollows where the eyes are lodged (one of many correlations Brook
creates between Lear and Gloucester). Then we see Tom from Lear's perspective, the camera tilting down to expose in close-up the beggar's vulnerable,
shivering, wet body while we listen on the soundtrack to Lear's speech on
"the thing itself."

When we finally enter the hovel the focus shifts away from that which
allows us to share Lear's subjective journey and goes instead to that which
causes us to observe the hero from the "outside" (a perspective closer to the
spatial context of Kozintsev's film). The shooting of the action that follows
gives us distance, a distance that shocks us into an awareness of the extent to
which we have been riding Lear's experience the whole way through. In
addition, when Brook finally gives us the "reprieve" of the hovel, we remember, in retrospect, many subtle but ongoing techniques of distancing
throughout the storm scene. Shots of Regan, Goneril, and Cornwall interrupt the sequence of Lear driving his horses, as do quick shots of the
frightened Fool inside the carriage. The juxtaposition of perspectives of
Lear and the Fool is a critical distancing technique for both Brook and

Shakespeare. As the filmmaker points out: "two views intermingle, the ferocious images that Lear sees, the equally frightening but more prosaic viewpoint of the Fool, for whom the storm is nothing but wind and water and cold."[8] The two perspectives create a dynamic in the imagination of the spectator that, ironically, enriches the power of the film's imitation of subjectivity by continually commenting on it. Similar moments of distancing occur when Lear first discovers Kent in the storm and when he sends the Fool and Kent into the hovel. All these moments pull us out of the immediate experience of the hero's crazed psyche, with the paradoxical effect of enhancing that experience through the technique of contrast. Then, when we enter the hovel, the perspectives of those around Lear seem to take over as the major guiding force. With the Fool we see a joint stool that Lear takes for Goneril; with Kent and Poor Tom we see sheep that Lear takes for the dogs Tray, Blanche, and Sweetheart. We observe from "outside" (relative to the sequence that precedes this) and watch the King conduct his trial. But when we share Lear's hallucinations in the hovel—hallucinatory visions of Goneril, Regan, and Cordelia—we recognize that Brook has simply shifted the balance. Instead of interrupting the guiding force of Lear's subjective point of view with perspectives external to it, the director interrupts the external points of view with the subjective visions of the hero.

An examination of the storm on film is particularly instructive because it is the sequence in Shakespeare's drama that epitomizes how the playwright creates his spatial field through the languages of the stage he worked with. The storm Shakespeare writes for the stage emerges through the imaginative activity that comes from the spoken word and the physicalization of the actor, in conjunction, perhaps, with a few sounds from backstage. (In a modern production, one could add the effect of lights, recorded sound, and scenery.) One can learn how the play operates by examining the conventions the spectator works with in the activity of observation. Shakespeare's theater functions through precise relationships among the elements of the stage that take form in the visual and aural fields and that operate according to conventions of time and space. With the conventions of his medium, Shakespeare "paints" a storm that transforms the neutral Elizabethan platform stage. In fact, from the earliest writing on the topic of Shakespeare and film, critics continually warn of the danger of a collision (or redundancy) of spoken language and cinematic images. Even Bert States, in his work on the phenomenology of the theater, expresses concern about film as a place "to speak [Shakespeare's] poetry in a milieu that usurps its descriptive function." He asks pertinent questions about the marriage of Shakespeare's language and the visual language of the screen: "What happens when our two fundamental forms of scenery collide at the pitch of their unique powers? What happens when a dense metaphorical world collides with a dense real world (real, of course, only in the sense of the explicitness of photography)?"[9]

States's point is well taken. If film reduces Shakespeare's poetry to a redundant visual description, then it does to the play something similar to what Lamb and Bradley claimed the theater did to it; in one instance, the crude languages of the theater destroy the playwright's poetry, and in the other, the highly developed language of the cinema collides with "a dense metaphorical world." True, both film and theater can butcher Shakespeare. But just as one might answer the earlier critics by stressing that the creative capacity of the theater is not to expropriate imaginative activity but to stimulate it in a space that allows the poetry to breathe, so, too, can one answer the concerns about film. If film language collides with Shakespeare's poetry, the result is unfortunate. If the film, however, offers a new context in which one can perceive the action of the drama, if the filmmaker can exploit the potential of cinema to place the language in a new space, a space where it sounds a little different to the ear precisely because it appears so different to the eye, then it achieves its maximum creative potential.

CHAPTER 4

Expanding Secrets:
The Space of the Close-up

Mr. Griffith turned to a young actor. . . . "Let's see some distrust on your face."

The young actor obliged.

"That's good!" Griffith exclaimed. "Everyone will understand it."

Billy Bitzer objected, as he was to do often when Griffith attempted something new: "But he's too far away from the camera. His expression won't show up on the film."

"Let's get closer to him then. Let's move the camera."

"Mr. Griffith, that's impossible! Believe me you can't move the camera. You'll cut off his feet—and the background will be out of focus."

"Get it, Billy," Griffith ordered. . . .

After the rushes were viewed, Griffith was summoned to the front office. Henry Marvin was furious. "We pay for the whole actor, Mr. Griffith. We want to see *all of him*."[1]

I

The space of the close-up is the space in which the tiniest detail can achieve the same magnitude as the largest and most imposing object; it is the space that reveals the "hidden mainsprings of life,"[2] a space of surprise, of unexpected action, of the details imperceptible to the everyday glance. But, as is evident from Henry Marvin's reaction, the close-up is also a strange space that we have, in time, had to adjust to; Marvin did not yet understand the power of the shot, that it does indeed allow us to see to an extraordinary degree, and that Griffith was not, in fact, cheating his audience by not showing "the whole actor."

The close-up shot in Shakespeare films takes us into a new space for the plays and adds an element to the performance dynamic the spectator must work with in the act of viewing. Welles's Falstaff, for example, that "huge hill of flesh," is suddenly, within the context of film, an aggregate of details—we see "*all of him*" in a very different way. Our reading of him is

Orson Welles's *Chimes at Midnight*. The "silent soliloquy" of Falstaff's face.

necessarily made up of a response to his face, in close-up with the warm smile in his eyes, the lines and crevices of his skin, and the layers of fat that articulate, in one glance, the story of a complex human being. With the close-up, Falstaff's face is as much a part of the *mise-en-scène* as Henry's castle, the Boar's Head tavern, the field of Shrewsbury, and the cathedral of Henry V's coronation. The material of Falstaff in close-up reveals his multiplicity. He is the nervous hand dismissing Shallow's memories of youth, the hand that holds in delightful celebration a cup of sherris-sack, the hand that takes bribes from the poorest of men, the hand that reaches for Hal's friendship, the hand of arrogance, and the hand of need. He is the mouth that smiles in wit and play, the mouth that quivers in embarrassment and vulnerability, the mouth that snores in drunken sleep, and the mouth that speaks to give a rhythmic center to the film. His eyes offer warmth and fear, languor and vivid energy, a longing for life and an acceptance of age. We understand Falstaff's character in the film by means of the close-up, the shot that offers, in the words of Walter Benjamin, "entirely new structural formations of the subject," the perspective in which "space expands." "The camera introduces us to unconscious optics as does psychoanalysis to unconscious impulses."[3]

Of particular significance is the capacity of the close shot to expose the human face in detailed magnification and to create a performance context in which the actor's physicalization of the role has much to do with his or her work with facial expression. Acting in film is, often, acting from the face; consequently, the viewer relies on the face as the map that charts the emotional, psychological, and spiritual journeys of a character. The film theorist Bela Balazs writes eloquently about the power of the human face in close-up (what he calls the "silent soliloquy") and stresses the creative processes it activates. "When we see the face of things, we do what the ancients did in creating *gods* in man's image and breathing a human soul into them. The close-ups of the film are the creative instruments of this mighty visual anthropomorphism."[4] The close-up achieves power through its capacity to stimulate the imagination of the observer, to engender the creative activity Gombrich discusses as "the beholder's share," or that Meyerhold teaches in his concept of the spectator as the "fourth creator."[5] The close-up changes our normal perceptual relationships and, in the case of the shot of the face, places us "in another dimension: that of physiognomy."[6] The active viewer now reads, in pictorial isolation, the emotions and struggles of another human being in the lines of the face, in the eyes, smiles, blemishes of skin, and wrinkles of age. The spectator enters a new "space" of human activity and experiences the world within a selective visual field. Human emotions (of both character and spectator) become a function of the unique spatial field of the cinema.

The 1969 version of *Hamlet* directed by Tony Richardson (with Nicol Williamson) is an example of a film that uses, almost exclusively, the close-up and medium shot.[7] Richardson's film warrants brief attention because it is a good illustration of how the close-up on the face works in a dialectical collage to form the *mise-en-scène*. Richardson reduces almost everything in this film to two elements: the spoken word and the face (or torso and face) shot against a black and undefined background. The characters talk about an Elsinore never realized in terms of a traditional *mise-en-scène*, an Elsinore that is a product of only faces and words. Gone are the repressive iron and stone of Kozintsev's film. Gone is the labyrinthine setting of Olivier's film. Gone is any material element in the setting that might reflect Denmark. In Richardson's film, the prison that is Elsinore is a function of cinematic form, of the tight spatial field of his shots, filled with the faces that crowd it. (Even when Richardson does take one of his few long shots, the persistent black background makes normal spatial orientation impossible.)

Richardson directs Act 1, scene 2, for example, as a scene of celebration and merrymaking for Claudius, his new queen, and the members of the court. He uses a collage of faces of people laughing, drinking, and feasting, faces that appear slightly grotesque in the magnified field of the close-up. We see Claudius and Gertrude, in alternating shots, eyeing one another

Tony Richardson's *Hamlet*. This still of Hamlet and Ophelia is typical of the tight shots used throughout the film.

lasciviously and drinking to their marriage. Wine dribbles down the face of Claudius. Laughter is everywhere and is the dominant force in this collage of faces in decadent revelry. And then Hamlet's face with its sorrowful expression bursts through to prick the bubble of high-spiritedness, a part of the collage that collides with the whole, a face of sadness among many of joy. His soliloquy ("O that this too too solid flesh would melt") finds expression as much through the pain and confusion he registers on his face as it does through his words. Shakespeare's play achieves a kind of Beckettian attenuation in this film with its isolated shot of the speaking figure and with its focus on a stream of words coming from a human mouth. It is a performance with the same paradoxical result as Beckett's plays: its very minimalism is a source of dramatic power. In the words of André Bazin, we witness "drama through the microscope" in which "the whole of nature palpitates beneath every pore. The movement of a wrinkle, the pursing of a lip are seismic shocks and the flow of tides, the flux and reflux of the human epidermis."[8]

Richardson's film illustrates the way the close-up operates through myriad locations of a given scene and that a montage of faces can serve, in dialectical relationships, as the total field of action. The spectator works with the fragments of many faces to complete imaginatively what the film, in its pieces, suggests. The close-up provides (to borrow again from Bachelard) an "intimate immensity," bringing the spectator into close range with what in the everyday world seems small but on the screen looms large. This change in relationship affects specifically what we see in the context of performance and particularizes the raw material of the filmic experience. The close-up, in its "intimate immensity," brings into relief the face that has always remained in the relative distance of the traditional *mise-en-scène* and forces it to stand up to intense scrutiny.

II

The framing of space in close-up is a technique that operates to create a sense of seclusion, of confidentiality, and, often, of secrecy. In bringing to the fore what is normally hidden or part of a larger whole, the close-up implicitly communicates to the viewer that he or she is somehow privileged, that he or she can look at a world normally unseen, or seen only within the context of a more comprehensive picture. The close-up is the performing space where secrets live, the isolated realm that gives expression to the clandestine, the undisclosed, and the cryptic actions of the plot. I do not mean to suggest here that the film somehow exposes new turns of the plot that come out of clandestine activity. I mean that the "secrets" of Shakespeare's plays find a new context for their realization. Covert acts find in the close-up an appropriate space for performance. It is the space where we observe, in isolation, the vial of poison poured in the goblet, the naked point of the rapier, the hands of a murderer, or the stabbing of a king.

Kozintsev's *Hamlet* illustrates the fact that the new, secret spaces into which the camera can take us are potentially dangerous (in a political sense). In the "Mousetrap" scene, for example, Kozintsev uses his camera to interrogate Claudius and Gertrude during the Player King's words:

> Our wills and fates do so contrary run
> That our devices still are overthrown;
> Our thoughts are ours, their ends none of our own.
> (3.2.209–211)

Kozintsev offers the spectator the chance to scrutinize the reactions of the King and Queen during this speech, and the subtleties imperceptible to the audience at a stage production become the entire dramatic focus. Responding to the penetrating words of the Player King, Claudius nervously takes Gertrude's hand. He then immediately lets it go on the words "So think thou wilt no second husband wed." With a focus on the sweaty palm and nervous twitch of the King, the field of action becomes the murderer's hand:

the "device" of political corruption and sexual degeneration. Through the close-up on this moment, our perception is exclusively directed to a metonymic image of the murderer and usurper, an image that makes visible in isolation what would otherwise remain hidden or camouflaged. Obviously, in live production we, together with Hamlet, catch the conscience of the King; in the cinema, however, the process takes place in the specifically filmic way of perceiving in a new spatial field.

In his film of *Macbeth*, Roman Polanski uses the close-up to create the space of strange rituals, of conspiracy, treason, scheming, and power politics. Like Shakespeare's own exploration of the tension between the quest for power and personal anguish, however, the filmmaker ultimately demonstrates that the space of conspiratorial activity is also the space of psychic torture. *Macbeth* is a play with a hero who, unlike Iago or Richard III, experiences not one moment of pleasure from his deeds. Every step he takes in achieving power has a cost, and Macbeth is keenly aware of the price he pays. Though the close-up gives to Macbeth's journey a context for the secret planning of "the Imperial theme," it nonetheless provides the place of the hero's "horrible imaginings"; the tension between these two products of the mind becomes the center of Polanski's film. The close-up is a mirror of subjectivity in this *Macbeth*, a mirror that reflects, in a tight spatial field, the suffocating world of the tortured conspirator, where conscience creates walls of constriction.

But Polanski does not stop with this technique of the visual field to create his effect. He exploits the voice-over to build in the aural field what the close-up achieves in the visual. Indeed, Polanski employs voice-over in this film more frequently than any other major cinematic adaptation of Shakespearean drama (it is used on approximately thirty separate occasions) to construct, through sound, a private space that exposes secrets. In considering Orson Welles's *Macbeth*, I demonstrated how the voice-over technique functioned in the dagger soliloquy, how it led to an identification with the hero because it allowed us to listen to the voices of his consciousness in a manner parallel to the way we listen to the voices within ourselves. The voice-over is a technique that imitates our own process of apprehending thought, and, because it places the action in a space different from the visual, it can often stimulate a creative dynamic among elements of a dramatic moment in new relationships. To add to that notion: the voice-over, as a convention of internal thought, is an aural image of the private; it is the sonorous realm of the undisclosed self. As Kaja Silverman points out, the voice-over "functions like a searchlight suddenly turned upon a character's thoughts; it makes audible what is ostensibly inaudible, transforming the private into the public."[9]

One of Polanski's greatest achievements in the film is the way our under-

standing of Macbeth's and Lady Macbeth's psychological journeys is a
direct function of this private spatial field, a field he creates by making the
voice-over the companion of the ubiquitous close-up. Ultimately, their experience is one of imprisonment in the scheming of their own imaginations.
Voice-over and close-up combine in the film to create, in spatial terms (both
visual and aural), the smothering prison of the mind; Macbeth tyrannizes
the world as he is tyrannized by the mind. And at the moment of Macbeth's
death in the film he is, appropriately, beheaded: a simultaneous release for
the suffering of Macbeth and the world that he shook.

From the very first sequence of the film, Polanski creates a sense of covert
and mysterious action through the pictorial isolation of the close-up. The
director disturbs the first long shot—of a barren and austere tidal flat at
sunrise—with a crooked stick that appears, in close-up, from the top right
corner of the frame, bisecting the picture and creating the first image of
severing in the film. On one level, therefore, the division is of the width (the
frame) of the shot. But the bisection is an act that takes place within the
close spatial field, an act that imposes on and changes, through contrast, our
perception of what is distant because it gives the picture a clear foreground
and background. It is thus a division of depth as well. In its divisiveness,
moreover, the picture of the crooked stick in close-up engenders a sense of
covert activity by defining that space in which that activity will find realization. The stick is a prop of the witches, who are about to enact a ritual that
serves as the filter through which we perceive the events of the film, a filter
of secret and strange deeds. The initial image announces the tight focus of
the frame as the space for strange enactments. It is a shot that anticipates the
division of Macbeth's own experience between the "Imperial theme" and his
"horrible imaginings," an experience of a split psyche. The appearance of
the stick begins a ritual as significant to the viewing audience as it will be to
Macbeth himself.

The witch holding the stick immediately begins to draw a circle in the
mud and, as she finishes, the arms of a second witch enter the frame (again
from the upper right hand side). The second witch begins to dig a hole with
her hands and is joined by a third woman, whose hands appear from the left
side of the frame. One witch is young and "fair," while the others are old,
withered and "foul" (yet another articulation of the visual motif of divisiveness in the film, relating, on many levels, to the notion of "equivocation" in
the play, a notion made explicit by the Porter). The ritual continues in close-
up, giving it a clandestine flavor, with a shot of one of the witches holding a
hangman's noose, followed by a full-frame shot composed exclusively of the
hands of the woman placing the noose in the hole in the ground. Polanski
follows that shot with a sequence of close-ups of (1) the faces of the three
witches huddled together; (2) a severed arm; (3) the arm, with a dagger fixed
in the hand, placed over the noose in the hole; (4) the hands of the witches

Roman Polanski's *Macbeth*. The witch's stick bisects the picture to create the first image of severing in the film.

Roman Polanski's *Macbeth*. The close-up as the space for the opening ritual of the witches.

burying the artifacts; (5) the hands of one witch pouring blood over the covered hole, (6) the witches huddled and chanting the last lines of Act 1, scene 1 ("Fair is foul and foul is fair. / Hover through the fog and filthy air"). The Weird Sisters then walk off into the distance. Two walk to the right and one to the left in another visual statement of division. The camera lingers on them for an exceptionally long time, and we observe them leave the foreground and diminish in size with every step they take until they disappear in the fog. They depart and will "meet again" in the "close-up" of Macbeth's imagination.

The entire field of action in the opening moments of Polanski's film is made up of hands, weapons of murder, barren ground, faces in profile, severed pictures—all combining in a sequence of a mysterious and cryptic ritual. In a sense, the same could be said for the film as a whole. The close-up exposes the face of the disturbed Macbeth when Duncan names Malcolm Prince of Cumberland and heir to the throne; it is the private space that exposes the sensuality of Macbeth and Lady Macbeth, the space of both their conspiratorial whispers and their kisses, a blending of elements central to Polanski's interpretive strategy. The close-up is the space of Lady Macbeth's conspiracy with the gods in her "unsex me" soliloquy; it is the space of the doubt in Macbeth's face when he decides not to murder Duncan, of his confusion while his wife goads him on, of his jealousy of Malcolm, who high-handedly commands Macbeth to fill his goblet with wine, of his panic before the regicide, of his dagger, of his bloody hands, of the murder itself. At a moment during the scene following the murder of Duncan, Polanski has Lady Macbeth grab her husband's arm to lead him to the well to wash the blood from his hands. What she pulls forward, however, and what the spectator sees in close-up, are the bloody daggers seemingly stuck to the murderer's hands. In the shot itself we observe the weapons, fresh from the killing, in front of Lady Macbeth's face. Like the crooked stick of the witches that enters the frame in the opening, the daggers in Macbeth's hand, slicing across the frame, effectively sever in fragments the picture of Lady Macbeth's face (a second way these weapons slash). It also exposes, in a shot of her shattered countenance, the horror she feels, horror that will gestate inside until it grows to overwhelm her with the madness of guilt.

It is in the moments when Polanski combines the close-up with the voice-over, however, that he most effectively realizes the smothering prison of the ambitious and tortured mind. The filmmaker first uses the combination in the sequence after Macbeth confronts the witches. The director delays the entrance of Angus and Ross until after Macbeth and Banquo have moved on past the stony crag where they first saw the Weird Sisters. Lying on his back inside a tent, eyes wide open and steeped in thought, Macbeth wonders about his experience: "The Thane of Cawdor lives, / A prosperous gentleman; and to be king / Stands not within the prospect of belief, / No more than to be Cawdor."[10] The camera at this moment shows the face of a man bewildered and intrigued, not believing the tale of the Weird Sisters, yet haunted by it, wanting to believe what seems incredible. Then Polanski cuts to a shot outside the tent for the arrival of Ross and Angus, who present Macbeth (as he goes out to greet them) with the title of Cawdor and a medallion sent by the King. During a close-up of the hero receiving the medallion, we hear his aside in voice-over: "Glamis, and Thane of Cawdor! / The greatest is behind." What was curiosity and intrigue in his facial expression and tone of (internal) voice is now spiced with the flavor of delight and a sense of satisfaction. The next cut—after a few words spoken by Banquo, the nobles, and Macbeth—is again inside the tent, where we observe in close-up the new thane, alone, placing the medallion around his neck. The shot tightens on his face as we listen to an internal monologue that increasingly voices the thoughts of a man giving rein to his ambition. "Horrible imaginings" enter the hero's mind at exactly the moment he puts on the medallion of Cawdor. On stage, we key into his thoughts through asides and soliloquies; in film, the voice-over and close-up work specifically to make those thoughts fill the entire space. On the screen the action *is* thinking. In the surreptitious space of voice-over and close-up, where the "murder yet is but fantastical," a conspiring imagination gives birth to the notion of regicide. Shakespeare suggests that Macbeth is so rapt by his imagination "that function / Is smother'd in surmise. . . ." In Polanski's film, thoughts swirl within the tiny space of the close-up and "smother" the man we see.

The development of Lady Macbeth is also a function of the close-up and voice-over combined. She reads her husband's letter out loud and then begins to delight, in voice-over, in fantastical imaginings—and the concomitant worries that her husband is "too full o' th' milk of human kindness." Her private scheming finds expression in the invisible space of the voice-over, while the close-up exposes such secret activities (relatively innocent at this time) as hiding the letter in a box in her chambers. When Macbeth arrives at Inverness, what was in the secret realm of the voice-over now finds expression in the encounter of husband and wife: the internal monologue of conspiracy in voice-over becomes the whispered dialogue of collusion. The camera keeps a tight shot on them as he carries her up the stairs during the opening dialogue. They then talk of Duncan and conspire to realize the claims of the Weird Sisters in a close-up shot composed of whispers, kisses, giggles, and caresses, a shot that articulates, by letting us in on their intimate encounter, the eroticism of their drive for power. As the film progresses, that erotic energy is smothered in the tight frame of a world of ineluctable and haunting thoughts that leave no room for the breath of tenderness and love.

Before Duncan arrives at Inverness, we see a close shot of Macbeth dressing and preparing for the event of the King's arrival—the medallion of Cawdor around his neck notably in view. From the right side of the frame

Roman Polanski's *Macbeth*. The close-up is the space of the secret whispers and kisses of Macbeth and Lady Macbeth, a space of conspiracy and eros.

his sword appears, just as the crooked stick of the witches appeared in the opening shot. Who gives Macbeth his sword? Presumably his wife, though we are never told in Polanski's visuals. Daggers "appear" before Macbeth just as weapons appear before the spectator and sever the picture on the screen. The King arrives at Inverness, and the dominant shot is a close-up of Lady Macbeth smiling and greeting her guest: the serpent looking "like th' innocent flower." The accompanying shot is a close-up of Macbeth himself, eavesdropping, hiding behind a wall in a place for spies, a place in which he finds nourishment for his incipient regicide. Like the moving camera in the static love contest in Kozintsev's *King Lear*, Polanski's visuals follow a subtextual line, and the film gives form to the hidden stirrings below the ostensible action.

In the raucous banquet given for Duncan that evening, Macbeth sits "rapt," locked in the struggles with a moral consciousness, a consciousness he eventually loses through the course of this film. The camera finds Macbeth's face amid the celebrations, and the voice-over finds his thoughts: "If it were done when 'tis done, then 'twere well / It were done quickly." Significantly, Polanski has all other sounds of the banquet almost fade out completely so that in the close-up and voice-over we enter an isolated realm of the hero's experience and respond to his dilemma in a performance space of exclusivity. Polanski's choice to set the monologue "within" the banquet attests to the power of the close-up to create a space of private contemplation. That the filmmaker maintains a faint buzz of the background noise (the most important element of which is Duncan's joyful—though for us at this moment muted—outburst, which the camera momentarily shows as well) gives the isolated focus on the hero a context for his deliberation. Though Polanski realizes the private realm of Shakespeare's drama, he does not do so at the expense of the relationship between Macbeth's thoughts and the world that suffers from actions born of those thoughts.

As Macbeth's doubts continue, we follow him outside to a porch, where he leans against a railing, watching the storm that has been raging all evening. The camera skulks around him, now showing him in profile, now from the back of the head. As we look at him from behind, we hear Lady Macbeth's voice in the offscreen space in back of the camera: "Why have you left the chamber?" With a shot exclusively on Macbeth, that voice startles us as it startles him. It comes from an invisible space; it interrupts the decision to "proceed no further." As an offscreen voice from behind, moreover, it functions in the same manner as the voice-over itself. In this case we do not see the face of the speaker but we know the voice well, a voice that collides with the words of the hero (words that are themselves exclusive to the aural field), who fears "deep damnation." In the following close-up shots she gently cries when her husband tells her that he will not do the deed. They go back inside the chambers to watch the entertainment of the evening, which includes a sword dance that expresses the kind of daring that Lady Macbeth fears her husband lacks. They whisper together as the dance ensues; we share in the intimacy of their interaction through the tight spatial field that, like the secret discussion, separates itself from all that surrounds it. When Polanski shoots the dance from the Macbeths' perspective, their whispering continues on the soundtrack. The aural field takes over in maintaining what the close-up provided as the space of private interchange.

Polanski's extensive use of the close-up to provide a space for the covert activity of the Macbeths continues right through to the visualization of the murder of Duncan.[11] The events leading to the regicide are made up of close or medium-close pictures that include Lady Macbeth's telling her husband that she will drug Duncan's guards, her smiling and dancing with the King, Macbeth's watching them dance and "thinking," in voice-over, that she should "bring forth men children only." In the dagger soliloquy, Polanski attempts to create a visual and aural dynamic that follows the waves of Macbeth's experience at that moment. By doing so, he momentarily de-familiarizes the private space he constructs throughout the film, thereby attesting to its power. Unlike Welles, Polanski decides to visualize the imaginary dagger and demonstrates its "invisibility" by having Macbeth's hand pass right through when he tries to grasp it. To articulate the swings of the hero's experience, Polanski has the dagger disappear and has Macbeth suddenly use direct speech on the line "Mine eyes are made the fool o' th' other senses / Or else worth all the rest." At that moment Polanski breaks the spell of private struggle created by the combination of spectral dagger and voice-over to parallel the moment Macbeth gains psychological footing and bursts out of the prison of his haunted consciousness. There is a sense of release in the direct speech, a purging of the internal voices that have begun to control the hero and that have filled the tight space of the voice-over.

But the direct speech quickly comes to an end, the imaginary dagger reappears, and Macbeth moves with it to Duncan's chambers. He passes the drugged guards (whom we see in close-up), finds Duncan, and freezes momentarily. The immediate prelude to the murder is a shot/reverse-shot sequence of the faces of Macbeth and the sleeping Duncan. The spectator experiences a dynamic of the horrified and, for a moment, reluctant hero (we take some solace in his momentary hesitation) and the innocent, sleeping King. In this film, we see Duncan wake and utter, in whispered surprise, "Macbeth." At that instant, we witness, in the detail of the close-up, the violence and bloodshed of the murder, violence we have come to associate with Polanski's film.

Polanski's bloodbath is obviously part of the adaptation process, part of how he revises the text for cinematic presentation and, in this case, is a deeply personal statement that comes out of the director's own experience. Though the questions of fidelity to Shakespeare are interesting to address, what is pertinent here is the space the violence occupies. The murders of Duncan, of Banquo, of Macduff's wife and children, and of Macbeth himself—all, at some point, involve a tight close-up shot, a shot that takes the murderer's hand, the victim's pain, the instrument of destruction, or the blood itself, and makes it the entire field of action. Both the scheming and the results of that scheming are, in the film, functions of the close spaces (aural and visual) made available by the medium. What Polanski's film does teach us about Shakespeare's play is that—unlike, say, the blinding of Gloucester, Hamlet's killing of Claudius, the murders of Caesar or Coriolanus—violence in *Macbeth*, until the death of the hero, never takes place in a public forum; violent acts in this tragedy are acts of the secretive and private realm. The spectator must then take all these isolated images and sounds of Polanski's film, all the hidden thoughts and acts of the close-up and voice-over, and organize them to build the film as a whole. The secrets and covert acts of the film, articulated in a unique spatial realm, are the raw materials that combine in the imagination of the spectator to create the film's texture and to generate meaning.

Following the murder, the close-up and voice-over, in addition to being spaces of conspiracy, are also spaces of guilt and psychological pain. Together they form the space that exposes Macbeth's fear at having "murdered sleep," as well as the space of the cover-up, epitomized by the full-frame shot of the hero washing his bloody hands in a bucket of water. The close shot is the place of whispers between Donalbain and Malcolm when they resolve to leave; it is also the place where we encounter, in the story of pain told by the hero's face, his defiled mind, his sleepless fits, his lost innocence (his face actually changes through the course of the film from beardless ingenuousness to a look of hardened, aging, coldheartedness—an evolution from fair to foul). In the space of the close-up, Macbeth and the two murderers conspire to kill the "enemies" Banquo and Fleance. (The shot is of three heads huddled together in covert activity, and is reminiscent of the full-frame shot of the witches themselves.)

But Macbeth has nightmares in this film and sees the ghost of Banquo. The bloody actions that were a product of a conspiring mind now become locked within the closed world of his psyche. To put it in spatial terms, Macbeth's greatest terrors become the pictures he sees in the "close-up" of his own imagination; Banquo's ghost is a product of the mind, a mind that once conspired to gain power but now suffers in a self-conspiratorial act of hallucination. The close-up reveals, with equal power, the dynamic of destructive forces operative both within and without; Polanski's *Macbeth* is a tragedy of the colliding forces of the mind unleashing the most brutal acts of violence the human being is capable of. The collision is expressed by way of the close-up.

When Macduff enters the fortress at the end of the film, he enters an isolated and lonely world from which all inhabitants have fled, a world populated only by Macbeth and the corpse of his wife. Indeed, towards the end of the film, the whole of Dunsinane (though no longer represented through close shots) becomes the lonely, private, isolated space it has always been in close-up: the private space of Macbeth's experience. At the moment when Macduff finally locates his enemy, Polanski takes the first long shot of Macbeth—tiny, alone, sitting in a corner, on his throne, like a child without a companion. In the distance, we can barely see him. The figure whose presence loomed so large throughout the film is finally rendered helpless in the space of the distant shot. Interestingly, we observe Macbeth in the distance from just behind Macduff, whose sword appears in the bottom of the frame in a shot that echoes the crooked stick of the witches, the sword of the hero when he dresses for the banquet, and the weapons of regicide that Lady Macbeth discovers locked in her husband's hands. The shot showing Macbeth's death is brutal and violent. We watch Macduff slice off his head—the part of the hero that served in the film as the dominant object of our observation. Macbeth is cut off from the source of his wretchedness: the mind that suggested the most "horrid image." Polanski's film takes place in the space of that mind and other minds that conspire with and give nourishment to it. Macbeth dies and is beheaded, a fitting end to a hero trapped in the imposing energy of his own imagination and guided by the conspiracy in the tight space of his mind.

III

The close-up shot has a number of elements that operate according to the laws of their own internal dynamic. Movement among the elements of a given shot ultimately participates in the movement of the film as a whole. Specifically, in a process similar to that of montage itself, the close-up, or medium close shot, can give a sociopolitical meaning to the drama by way of the dialectical field of the single, magnified frame. It strengthens in dramatic power key relationships that have a specifically social makeup, because it juxtaposes action with a character's attitude toward that action, told through the details of facial expression. In the rest of this chapter, I shall explore the tight shot as the site for a specifically political dimension of the Shakespeare film, one that is generated by the dynamic operative in the tableau of the close spatial field.

In a moment (almost fleeting) in Olivier's *Richard III* near the beginning of the scene in which Gloucester woos Lady Anne, the director takes a tight shot of Richard and Anne leaning over the corpse of Anne's husband Edward.[12] Those are the only three elements of the shot: widow, corpse, and murderer–wooer. At that moment the spatial field of action is, ironically, made infinitely more complex, even though the visual elements themselves are simple and direct. I have spoken earlier about a dynamic of perspectives that comes from editing and the effects of multiple angles on a given object. In this instance, however, the filmmaker is able to construct a dynamic within a single shot; we move, in other words, from an Eisensteinian notion

Laurence Olivier's *Richard III*. The tight shot of widow, corpse, and murderer-
wooer.

of shots in dialectical opposition to the "window" on the world of the single
shot hailed by André Bazin. But unlike Bazin's specification of deep-focus
photography and the elements that work in a dynamic through that technique, I wish instead to point out how the close shot itself, in its ability to
isolate dramatic action in a very precise way, works as a field of dialectical
activity. The dynamic Olivier offers in this tableau resonates for the entire
film; it articulates the most fundamental political design of the drama. It
offers a reference point for Richard's deeds, both past and future; it crystallizes how Richard works, how he changes faces, how he infects others—
in short, to borrow from the vocabulary of Bertolt Brecht, it articulates the
fundamental *gest* (*Grundgestus*—"basic *Gestus*") of the film. And it achieves all
this with the spatial field of the close shot.

   In the space of the tight shot in Olivier's film, we enter into the realm of
gest because we enter into a space of attitudes. Recall Brecht's words in "A
Short Organum for the Theatre":

> The realm of attitudes adopted by the characters towards one another is what
> we call the realm of gest. Physical attitude, tone of voice and facial expression
> are all determined by a social gest: the characters are cursing, flattering, in-
> structing one another, and so on. The attitudes which people adopt towards
> one another include even those attitudes which would appear to be quite
> private, such as the utterance of physical pain in an illness, or of religious faith.
> These expressions of a gest are usually highly complicated and contradictory,
> so that they cannot be rendered by any single word and the actor must take

care that in giving his image the necessary emphasis he does not lose anything, but emphasizes the entire complex.[13]

Conceptually, then, Brecht's notion of *gest* accommodates contradictions as it encapsulates the "attitudes" characters have toward one another, attitudes that are our clue to understanding social relations because they are informed by a specific sociopolitical context. It is a somewhat bewildering concept because the *gest* must be absolutely simple (the most rudimentary definition of *gest* is "gesture") while accommodating what is "highly complicated and contradictory" (a gesture that is packed with complex connotations). As Martin Esslin points out, *gest* is "the clear and stylized expression of the social behavior of human beings towards each other," and such human interaction is at the heart of Brecht's epic theater.[14] Brecht's notion is compelling (and one that attests to the expressive powers of the stage) because it teaches how the simplest sign in performance can resonate with the dynamic of social relations that inform it: a slave bows to his master, Mother Courage bites the coin she receives from the Sergeant (as the Recruiting Officer takes away her eldest son), Mrs. Peachum holds before Jenny a bag of money when plotting against Macheath, and so forth. In all these examples "the detail of the gesture has a political meaning, rendering properly and correctly the differing alienation of the roles. . . . meaning is no longer the actor's truth, but the political relationship of situations."[15]

The close shot in film, because of its capacity to render the smallest detail as the entire dramatic focus (a capacity to render as well the "private" attitudes Brecht discusses), and because of its potential to make the simplest image represent complex social and political relations, takes on *gestic* content. As the *gest* in Brecht's theater fundamentally has to do with human (political) relations, so Olivier's tight shot frames those relations in Shakespeare's play and the attitudes of the characters who constitute them. Like the positioning of characters on Brecht's stage in a central tableau, or like the simple gesture of his characters (which are as fleeting in live performance as the still picture in the film), Olivier's shot articulates through its composition, political and social meaning.

I should say a word at this point about the shot under analysis in *Richard III* only because it passes so quickly in the film. Though in the act of viewing the spectator could never dwell on the shot itself (it lasts maybe five seconds), I believe that in the slower time frame of contemplation and analysis in the study, one can unpack it to find the very center of Olivier's directions of the moment. Though it may seem that I am not looking at the close-up here in its natural state as part of the "moving picture," my goal is to achieve just that. In other words, I isolate the still to see if the "moving" elements that make it up can reflect on the "moving" pictures of the scene as a whole. Unlike my earlier examination of Polanski's film, concerned with what happens among many shots in close-up, I here follow Barthes' method of looking "*inside* the fragment, into the elements included in the image itself."[16] In addition, the imprint of fleeting images should not be underestimated; to have an isolated shot of Gloucester, Anne, and the corpse of her husband is compelling, to say the least, and gives the scene a specific context for the interrelations of its central elements.

Richard leans over the corpse of Edward towards Lady Anne, who kneels by the open casket in mourning. That is the simple shot. But it resonates with a sociopolitical level that is "highly complex and contradictory." Richard's character finds expression through the number of masks he wears simultaneously. He is at once, in the picture, contrite murderer (a mask he wears for Anne), fulfilled murderer (his act was a necessary step in his political climb), indifferent murderer (to us in the audience who have already heard him admit as much) and the murderer of sexuality between a man and a woman. He is the carrier of a new force of sexuality and imposes it as he invades the intimate space of mourning between a widow and her dead husband. He is at once the cause of blood of the past, the political machine of the present, and the victor of the future. The vampire leans over to suck out all vitality in a private space he steals to make his own.

Anne, too, articulates a complex of attitudes that includes the inexplicable crossing of feelings of repulsion and attraction, of hatred and desire, of a past that persists for her in the present (the corpse of her husband) and a future that mercilessly drags her forward (Richard). She is one of Richard's agents of revenge and a victim of his political intrigue; she is simultaneously powerful (in that Richard considers her a necessary step for his advance) and a cog in a political machine that destroys her identity both as woman and as one who once functioned within the filial power structure of England. Finally, the corpse of Edward recalls the history of destruction in the Wars of the Roses and the role of individuals such as Richard in that history. It is an image of the years of suffering and loss (back to Richard II) that inflict themselves upon the present. Edward's remains tell the story of brother against brother, of pain and loss; the corpse is the presence of a perverse history in the moment when murderer woos widow.

No one element of the shot takes precedence over another; the whole works as a complex of relationships activating the imagination of the viewer. The shot crystallizes the relationship of sexuality and death, suggesting the parallel between Richard's physical destruction of Edward through murder and his destruction of Anne's sexuality. The close shot becomes the dialectical field for imaginative activity. In the "intimate immensity" of this close realm of action, Richard woos; but the act of wooing functions in relationship to the expression on his face, which speaks of exploitation and power

politics. The multiple masks operate in tension with the act of pursuing
Lady Anne. Similarly, Anne's spitting in Richard's face is complicated by
her sexual attraction to him (something in her eyes). (Her attraction is
especially evident later in the scene as she seems transfixed by him, involun-
tarily staring at him and turning in his direction as she tries to leave.) All
details of expression, enormous on the screen, convey a realm of attitudes
that works in tension with the action itself to produce the social *gest*: "Richard leans towards Lady Anne over the corpse of her husband." This is an
act of political significance.

The social context that informs this moment, moreover, has to do with a
"house divided," with a severing in two of what can find strength only in
unity. There is a crack for a Richard to enter, created by the divisiveness of
England; only in division and civil strife does Richard find a world that he
·can bustle in. In the image, Richard's act of leaning over effectively creates a
gap (psychologically) between Anne and Edward, an action that simultaneously fills part of another gap between himself and the throne. Shakespeare takes care to show that Anne, as a product of a divided world, is
divided in herself, unable to summon the strength that would come from
being whole. This is not to suggest that Anne is a weak figure. On the
contrary, the scene could not play, or would play with little interest, if she
were not a match for him. She is, however, as vulnerable as England herself,
divided (the cause of which she cannot understand), mysteriously lost, split
in need and desire. Richard wins Anne as he wins the country, and Olivier
gives that action a concrete image in the tight shot. Richard takes over the
space of a widow and the corpse of her husband, the man he has murdered.
He infects that space, as he infects all of England. He finds the gap and
slithers in to take control and begin his work. Olivier gives the isolated close
shot *gestic* content.

Kozintsev also uses the tight spatial field to construct the social *gest* of his
film. He does this most effectively in memorable shots of Ophelia, whose
experience in the tragedy seems to encapsulate the filmmaker's emphasis on
broken relationships and alienation. One thinks of the medium close-up on
Ophelia during an early scene in which she dances rigidly to a tune an old
woman plays on the lute. The stilted and formal pose she takes, coupled
with the resistance to the dance we perceive in the grimace on her face,
made evident with a close-up, creates a tension articulating social *gest*: it is a
shot of imposed activity and the attitude of the character to that activity.
Another close-up shows Ophelia sitting obediently on the floor by her
father's chair, while he wags his finger in her face and drones on incessantly.
The shot, in its isolation of the wagging finger and the innocent face op-
pressed by that finger, crystallizes the social relationships (and filial politics)
that inform Denmark on many levels. But perhaps the most powerful of all

the close images is the one of Ophelia after her father's death when attendants in black force an iron corset and hoop onto her body (and later a black
dress of mourning), an image complete only when that symbol of repression
is juxtaposed with the pain of mourning in her face (whose tension the close
shot catches). The collision of her pain with this clear symbol of captivity
(the iron corset recalls the portcullis that imprisons Elsinore) parallels the
larger collision of the human spirit with the political regime and makes the
social *gest* clear.

The social *gest* evident in a close shot of Claudius in the prayer scene is
significant for showing another side of the tyrant, for letting the audience in
on an attitude exposed only by the tight frame. (In this particular shot we
move, with Kozintsev, from relations between people to one's relation to
oneself.) It is the picture that finally articulates Claudius's self-alienation
and shows that the tyrant is not exempt from the consequences of his own
political oppression. Instead of watching Claudius kneel down to pray (recall that Kozintsev de-emphasizes the Christian context of the play), we see
him confront himself in a mirror. Beginning with "What if this cursed hand
/ Were thicker than itself from brother's blood," the camera reveals a
medium-close shot of the King on one side of the frame and his image
reflected in the mirror on the other. The full-frame gives us a perspective on
three aspects of the moment simultaneously: the King, his reflection, and
the image of a man who is confronting his own reflection. As we perceive
the King in this multidimensional shot, he perceives himself in the "close-
up" of the mirror image; this "internal" image, the close-up *within* the film,
is a second text that parallels and participates in the dynamic of the first text
of the film. The King's ambivalent attitude to his crime is an ambivalence to
the image of the self; the act of confronting his own image is an act of
examining his deed (especially since it comes directly after the "Mousetrap"
sequence). In facing himself, Claudius faces the very source of disease in
Elsinore. The network of attitudes in the shot is highly complex, and the
social *gest* resonates in a tension of politics and lust, a tension Claudius
expresses with the words "My crown, my own ambition and my queen."
Again, the close-up functions as a space for the social *gest* of a scene because
it provides a context for the dialectical tension between act and attitude.
New possibilities for imaginative engagement open when the new relationships within the tight shot activate a dynamic that is unique to, and a
specific function of, cinematic space.

CHAPTER 5

Local Habitations: The Dialectics of Filmic and Theatrical Space

I

A major part of the achievement of film is the creation of a new spatial field for the realization of Shakespeare's plays. Whereas Shakespeare articulates space through words and action on an open platea stage, the filmmaker uses a technological language to create space through a distinct sense of movement, a dynamic of perspectives, details of *mise-en-scène*, sound, and the range and distance of shots. Shakespeare often comments on the nature of his own medium and its spatial attributes by informing his plays with a theatrical self-consciousness or metadramatic dimension.[1] Throughout his writing, he delights in the imaginative possibilities of the stage, in the process of turning the boards into a place of infinite variety, in the metaphorical richness of the theater as an image of life. To discuss Shakespeare's spatial field is to recognize, not only the practical element of painting the scene through words and actions, but also the conceptual exploration of the playwright as he muses on that process. In other words, Shakespeare creates space in his theater with a pen that "gives to airy nothing / A local habitation and a name," but does so in a spirit of critical inquiry about the nature of his art.

Operative in Shakespeare's plays, therefore, is often a relationship between creative activity and a self-consciousness about that activity, between a definition of the spatial field and a recognition of, or commentary on, the game we play with the poet in creating that definition. As the filmmaker constructs a space, he or she also has the capacity to bring attention to the cinematic art and to imitate, in that medium, Shakespeare's own self-conscious reflections.

It is Laurence Olivier, with his roots firmly in the tradition of the stage, whose Shakespeare films illustrate how the cinema can create a dynamic

Laurence Olivier's *Henry V.* The Archbishop of Canterbury and Bishop of Ely in the opening scene at the Globe.

between qualities of space, both theatrical and filmic. By so doing, the filmmaker builds his spatial design in a critical spirit patterned after Shakespeare's own explorations of the stage. In *Henry V,* for example, Shakespeare brings to the play a pronounced theatrical self-consciousness; Oliver imitates this with a filmic self-consciousness by the use of the spatial field he constructs. The critical discourse on Olivier's Shakespeare films abounds with discussions on relationships of things filmic and theatrical. What I wish to add here is the argument that the dynamic that Oliver creates provides a direct parallel to this self-referential, metatheatrical aspect of Shakespearean drama.

Imagine that "this wooden O" represents "the vasty fields of France." "Suppose within the girdle of these walls / Are now confined two mighty monarchies, / Whose high-upreared and abutting fronts / The perilous narrow ocean parts asunder. On your imaginary forces work." These words of the Chorus at the opening of Shakespeare's *Henry V* call more directly than in any other passage in the playwright's work for the imaginative participation of the spectator. The spatial field of Shakespeare's drama must operate through the imaginary forces of the audience. With an exploration of theatrical processes that informs the entire play, the playwright teaches us that his is a theater not of passive observance but of active participation. Olivier takes this statement as the starting place for his conceptual strategy. He frames the film with a simulation of the first performance in the Globe Playhouse on May 1, 1600. When the scene shifts to Southampton, he takes us into a second performance space by removing the action from the Globe world and placing it on a three-dimensional ship set against a painted, illusionistic backdrop. (It is a space, as Jorgens points out, reminiscent of "the illusionistic nineteenth-century stage.")[2] A third performance space materializes with the Battle of Agincourt, where Olivier comes closest to exploiting the cinema's capacity for photographic realism. In this third space, we are plucked out of the two-dimensional world of painted backdrops and cardboard flats and placed in a world distinctively filmic.

The opening of the film serves to bring out, in an ironic crossing of media, the bold reality of *Henry V* as theatrical entity; paradoxically, there is no pretense that what we are about to see, on film, is anything but a play in performance. The hybrid of theatrical and cinematic space operating at this moment is highly suggestive. The Chorus enters the stage to the applause of the Globe audience and delivers the prologue, an opening speech that not only calls attention to the reality of the "wooden O," but (in the context of cinema) emphasizes the essential activity of imaginative engagement in film as well. The dynamic of theatrical and filmic spatial fields is most fully realized at the moment when the Chorus says, "On your imaginary forces work." On that line the camera zooms to reveal the face of the speaker in close-up and suddenly, even shockingly, we are fully in the realm of cinematic space. The shift to close-up creates a sense that the Chorus has turned his attention away from the Globe audience and toward the film audience. We are suddenly aware of our place in the cinema, of our role as observers of one spatial field grafted onto another—all because the close-up makes us feel as if the Chorus is speaking directly to us. In other words, the spatial field of the close-up alienates us from our participation in the specifically theatrical space of the Globe world. Olivier thus uses the close-up to reflect on the film medium itself; with his camera, he parallels Shakespeare's technique—evident in the words of the Chorus—to make his audience aware of the nature of the performing space before them. In the film, however, the spectator has the special opportunity to cross the line between the spatial fields of stage and cinema to experience a rich blend of filmic and theatrical language.

The close-up on the Chorus for his line "On your imaginary forces work" crystallizes the relationship between theatrical and filmic space, and the moment serves as a paradigm for the dialectical interplay operative, at some point, in all of Olivier's Shakespeare films. We are aware of a scene or moment functioning according to the laws of stage space (albeit on the screen), only to be shocked out of that awareness with a sudden cut to a shot that is uniquely cinematic. The reverse happens as well. Through close-up, moving

Laurence Olivier's *Henry V.* Ancient Pistol enters the Globe stage—an example of the first (and most theatrical) performance space of the film.

camera, montage, and so forth, we are lodged in the world of cinematic space and then suddenly recognize, in the next shot, that Olivier is asking us to complete the scene through imaginative processes we associate with the stage. As a result of the interplay of the characteristics of these two "performing spaces," the filmmaker comments on his medium, on the medium the plays were written for, and, simultaneously, on the metadramatic element that informs the plays in the first place. Cinematic space accommodates qualities and characteristics of the stage (through a stationary camera, illusionistic backdrops, descriptive passages, etc.) as well as those of the screen, and, in the dynamic between them, places the action in a self-conscious spirit similar to Shakespeare's own meditations on the art of the theater.

Olivier transports us from the first performance space—the Globe—to the second, as Shakespeare transports us to Southampton. The Chorus again figures prominently in our journey as he delivers a truncated version of the opening of Act 2 beginning at line 31: "Linger your patience on, and we'll digest / The abuse of distance, force a play." While he speaks, he directs our attention to the curtain of the Globe stage. The camera—our perspective—slowly closes in on the curtain to reveal a painted rendering of the harbor of Southampton. Cinematic space layers itself on theatrical space. With a dissolve, Olivier takes us, "in motion of no less celerity / Than of thought" out of the Globe and into the self-consciously illusionistic space of phase two. What Olivier offers with the three-dimensional ship set against the painted backdrop resembles what the Chorus describes at the opening of Act 3; we see "the well-appointed King at Hampton pier / . . . ; and his brave fleet / With silken streamers the young Phoebus fanning."

Olivier thus substitutes a space of bold theatrics (with a specifically documentary flavor) with a world that is striking for its stylized perspectives and for its attempt to reproduce the historical period of Henry V. The costumes and backdrops are taken, in detail, from the medieval illustrations for the *Calendar of the Book of Hours* of the Duke of Berry. The bulk of the action that follows the Globe sequence unfolds in this second performance space. The viewer is keenly aware of the contrast between three-dimensional actors and the decorative compositions of the set; the flavor of theatrical artifice dominates. On another level, however, Olivier has placed us in a realm that is, relatively speaking, more cinematic than the Globe world because in this second phase the camera enjoys a freedom of movement not possible in the first setting. That movement is key to the definition of cinematic space. The mixture of qualities here is significant. By creating self-consciously illusionistic space that unfolds in a relatively heightened cinematic context, Olivier makes the audience continuously aware that they must "still be kind / And eke out [the] performance with [their] mind." Olivier maintains a level of illusion so that his film can act as the same kind of trigger for active imagining as Shakespeare's open stage. In Olivier's second phase of action, space expands cinematically, but the persistence of a theatrical backdrop always holds that expansion in check.

The shift from the second to the third performance space occurs with the Battle of Agincourt and operates according to the paradigm Olivier establishes with that key close-up on the Chorus. In a defamiliarizing relationship similar to the one of the close-up and the theatrical Globe world, phase three is a filmic space that sheds light on the theatrical qualities of phase two. Despite the controversy among critics of the film about the "realistic" nature of the Battle of Agincourt, it is undeniable that, within the conventions of this film, a stylistic shift occurs for these scenes. In other words, as Gombrich insists, what is significant is the relationships of the elements in a work of art; it is thus not pertinent here whether one can categorize the action of the battle, in an objective sense, as filmic realism. Only the internal relationships at work concern us.[3]

In his presentation of the battle, Olivier creates a sequence of colorful pageantry and visual detail. He shot the battle on location amid the green rolling hills of the estate of Lord Powerscourt at Enniskerry near Dublin.

Laurence Olivier's *Henry V*. This shot of Princess Katharine of France is an example of Olivier's second performance space, which clearly depicts the action against a painted, illusionistic backdrop.

Approximately five hundred men of the Eirean Home Guard and two hundred Irish horsemen from nearby farms played the French and English knights and soldiers.[4] The shift into the third phase begins with a dissolve from Henry's prayer at the end of Act 4, scene 1 (the night before the battle) to a full-frame close-up of the fleur-de-lys on a tent flap as it blows in the wind on this fateful day in history. The shots that follow reveal the preparations for battle on both sides. Through the composition of those shots,

Laurence Olivier's *Henry V:* before the Battle of Agincourt in the third performance space. Olivier plucks us out of the two-dimensional world of painted backdrops and cardboard flats and places the action in a world distinctively filmic.

Olivier realizes the disparity between the "lusty" and overconfident French on the one hand and the apprehensive and cautious English on the other. As the battle sequence ensues, we eventually realize that we are no longer in the artificial world of phase two, but have entered an expansive space of cinematic movement. In a manner echoed by Welles in his treatment of the battle of Shrewsbury, Olivier uses montage to create a rhythm of tension and excitement as the confrontation takes place. Unlike Welles, however, Olivier does not take us into the intimate space of pain in battle so much as he glorifies England's great triumph (central to Olivier's purpose at this historical moment). He does take a moment to linger on the vicious attack of the French on the young English boys guarding the tents and the subsequent grief and anger of the King and his men. But brilliant English tactics, the advanced use of the longbow, clever ambush, and the defeat of the many by the few are the points of focus in the space of the battle on film.

Just as the Battle of Agincourt represents the moment in the play when the imagination must accommodate the "vasty fields of France," just as the battle requires for its dramatic realization the imaginative expansion of space, so, too, does the battle on film represent the moment when the director parallels the activities of the imagination and expands his performance space to its maximum. Olivier imitates through cinematic technique the imaginative processes of the audience of Shakespeare's great epic. He accomplishes this through the dialectics of theatrical and filmic space.

II

Olivier works in a similar way in *Richard III*. He creates a spatial dynamic that comments on the theatrical self-consciousness of Shakespeare's play by using a parallel filmic self-consciousness. In this instance, however, Olivier reflects not so much the imaginative processes of live theater as the relationship of Richard, Duke of Gloucester, to the distinctly theatrical world he bustles in. Richard knows how to smile and play the villain—"I can smile, and murder while I smile." As I indicated earlier, Olivier builds a *mise-en-scène* with a specifically theatrical flavor to highlight Richard's role as the master of ceremonies, as one who stage-manages his rise to power by manipulating the world around him. England is Richard's stage. In addition (and more pertinent to the present discussion), Olivier juxtaposes qualities of stage space with characteristics that are uniquely filmic, in a dynamic that calls attention to the significance of the spatial field to the drama.

A telling example of Olivier's tactics is evident early in the film when Richard stages the imprisonment and ultimate death of Clarence. This sequence crystallizes the relationship between stage and filmic space and forms a paradigm for the film, just as the close-up on the Chorus at the

Globe formed a paradigm for *Henry V.* In the play, Richard tells the audience in his first soliloquy how he has "set [his] brother Clarence and the King / In deadly hate the one against the other." He reports how he has framed his brother George, Duke of Clarence, by informing King Edward of a treasonous plot based on "a prophecy, which says that G / of Edward's heirs the murderer shall be." Immediately following these lines, Clarence enters, guarded by Brakenbury, the Keeper of the Tower, and the episode that follows is the first glimpse we get of Gloucester's keen dissembling. He feigns love and concern for his brother's plight, blames the Queen, and promises Clarence that he will do all he can to ensure his release. Upon Clarence's exit, however, Richard turns to his audience with glee to tell of his plan to have his brother murdered.

In the film, on the other hand, the sequence begins as a play of shadows. We see, projected on the floor, the shadow of Richard approach the shadow of the King sitting on the throne. We do not hear what Richard says, but see, in shadow, that he whispers something in the King's ear.[5] The following shot shows in actuality what we just observed in shadow, displaying as well Edward's reaction of horror to Gloucester's message. Richard kisses Edward's hand and walks away. Olivier shoots the shadow of Richard as he departs. We are clearly in the realm of filmic space, but we already sense the interrelationship between stage and film. The isolation of shadows, while it is certainly, within the film, a convention of evil and of ominous activity, also provides a comment on the performance spaces operating. With the entire spatial field filled with shadows, we witness the suggestion of an event, a play of shadows that reflects at once on Richard as shadow and actor (perhaps Olivier is playing with the synonymous nature, in Elizabethan vocabulary, of these two terms?),[6] on the theatricality of his enterprise (the whole episode is clearly one Richard "stages"), and, curiously, on the play of shadows and light that is the film itself.

In the following moments, however, the filmmaker pulls us out of the filmic realm and places us firmly in the theatrical one. By doing so, Olivier highlights how Richard's whispered words turn quickly into action and how, like a playwright, he stages what he describes. After the play of shadows, Richard walks along the wall just outside the throne room telling us, his unwitting conspirators, of the plot he has laid. He stops when he reaches a window and opens the shutters to expose the events inside. The King charges Clarence with treason, and then faintly we hear the Duke's response in a cry of innocence. The man of the theater momentarily opens a window onto his stage to flaunt his handiwork. He closes the shutters, completes the remaining lines of his speech, moves to a second window, and opens it to exhibit the King's remanding his brother to the Tower. Richard closes the shutters of the second window as a stage-manager closes the curtain on a stage.

The significance of Shakespeare's theatrical self-consciousness in this play is inextricably linked to a pattern of description and deed, a pattern Olivier stresses through the spatial field he creates and one that would be of particular interest to Kenneth Burke.[7] The film shows the enactment of the evil plot as Richard describes it; we see Richard construct the scene before us while we simultaneously witness him taking pleasure in his creation and the political opportunities he develops for himself. By presenting the simultaneity of language and action, Olivier features the power of Richard's greatest weapon—words. The filmmaker demonstrates that Richard controls the space of the drama, that to describe is to enact, and that a parallel exists between Richard and his world and a playwright and the scene. Interestingly, Olivier excises the character of Margaret, and with her go the curses that help to structure Shakespeare's play—she serves in the original as a counterpart to Richard, as one whose words carry with them a force of enactment. In the film, Richard shares the power to write the script for the world with no one. When he finds his allies (most notably in Buckingham), he teaches them how to use his rhetoric to serve his purposes.

The layers of activity become most interesting when one recognizes that the juxtaposition of words and action parallels, not only the relationship of a playwright to a scene, but that of a filmmaker to his spatial field. As words are behind action for Shakespeare and Richard, theatrical space (giving the film a theatrical flavor) is behind that which determines overall filmic space—all techniques that open the spectator to an awareness of the nature of the medium in operation. The cinema's capacity to juxtapose qualities of both filmic and theatrical space exposes the processes of manipulation that layer the performance(s).[8] Shakespeare demonstrates to his stage audience that, as a playwright, he creates space with his power of language; Richard demonstrates to both the stage audience and the "cast" of characters he manipulates that England becomes what he *says* it will become; Olivier demonstrates to his film audience that what he shoots of Richard's activities in filmic space must be understood in the context of the stagelike world in which these events unfold.

Examples of Olivier's technique abound. The wooing of Anne begins in a long shot that exposes a distinctly theatrical space. Indeed, the funeral procession enters a space that the camera is already in; like an entrance on the stage, the scene comes to us, as opposed to the convention in film in which camera movement or editing "finds" a scene about to start or already in progress. The use of the long shot to offer a theatrical feel works in this film for many reasons. First, it is the shot that most closely resembles the stationary, panoramic perspective of live stage. Second, it is the shot that exposes most clearly the illusionistic background of Olivier's *mise-en-scène*. And, finally, Olivier constructs spatial relationships with actors, furniture, and props to form tableaux with a distinctive stage quality, tableaux evident only in the long shot. When, however, Olivier zooms in for a close-up on Richard, Anne, and the corpse of Edward, we are suddenly in a distinctly filmic world. The effect of this change in spatial relationship is to defamiliarize the theatrical space in which the scene began in order to emphasize, by contrast, the nature of the game at work. We are paradoxically more aware of the theatrical underpinnings of the scene by experiencing a spatial realm (the close-up) the stagelike world is in contrast with. The result is to give the wooing scene a spirit of theatrical contrivance. In addition, we are alienated from the filmic space of the close-up when Anne leaves and Richard speaks to us in soliloquy—a convention that carries the essence of theatricality and, in its own pattern of defamiliarization (a stage device on screen), calls attention to the film medium. In this way, Olivier reflects on cinema in a manner parallel to Shakespeare's reflections on the art of the stage.

Similarly, the theatrical self-consciousness of the play (and Richard's specifically theatrical role in it) is evident in the scene in which Richard and Buckingham conspire to "infer the bastardy of Edward's children." The scene unfolds in a juxtaposition of close-ups (filmic space) of the scheming dukes with long shots (theatrical space) of the little Prince playing on the throne. Later in that scene the camera remains back when exposing the events with Richard's little nephew York. The filmmaker challenges that sense of theatrical space, however, with a sudden close-up on the furious Richard when his nephew makes the seemingly innocent jab: "Uncle, my brother mocks both you and me: / Because that I am little, like an ape, / He thinks that you should bear me on your shoulders." Music swells forebodingly, a shot–reaction shot pattern shows the terrified York and the enraged Richard, and we are unmistakably in a filmic world. When Catesby leaves Richard and Buckingham to check on the political loyalty of Hastings, the villain stops his messenger just outside the throne room with his quip: "Give Mistress Shore one gentle kiss the more." Richard gently closes the door and the camera lingers on it momentarily, establishing a theatrical feel as we witness an image comparable to the closing of a stage curtain. The next cut shows, in close-up, Hastings and Shore in a passionate embrace. The action now is in the realm of filmic space.

Similar dynamics of stage and filmic space inform the scene in which Richard orders the execution of Hastings, or the scene of the death of Hastings himself, the murder of Clarence, the coronation of Richard, the conspiracy with Tyrrel, and the rejection of Buckingham—all calling attention to the theatricality of Richard's power. Significantly, after the conspiracy with Tyrrel is complete and Richard begins to lose power, Olivier remains increasingly in the realm of filmic space and calls less attention to the play's theatrical self-consciousness. In other words, as Richard loses control, he simultaneously loses his power with words (forgetting, for example, in Act 4, scene 4, to express the purpose of his order to Catesby to go to the Duke of Norfolk), and thus his power to write the script of England's history. Olivier takes away Richard's theatrical prowess by diminishing theatrical space.

Buckingham's participation with Richard is pertinent precisely because the former represents a continuum of the latter's power; he uses language in the same way Gloucester does and understands how to create the scenarios of his plot—a fact most evident at the moment when Richard, standing between two clergy in front of the citizens, agrees, with seeming reluctance, to ascend the throne. In that scene, Buckingham is Richard's fellow actor. Olivier begins the alliance of these two men during the scene in which Edward feebly tries to create harmony and peace among the nobility of the court. Olivier combines Act 2, scene 1 and Act, scene 2 to create one scene in which peace among the nobility is seemingly made in front of the ailing King. The King's wish to put an end to the enmity and factiousness of his court points out, by contrast, how poor a playwright he is compared to Richard. The moment in the film is a comic and unconvincing enactment of a "kiss and make up" game. Edward stages a scene, but his characters lack conviction; all present merely humor the ailing ruler. Moreover, when Richard enters the scene and claims that Edward's revocation of the order to execute Clarence came too late, we recognize the real playwright of the scene because Buckingham and Catesby stand by in silence. These two men know full well that Richard is lying, because, in the film, they were the ones to give Richard the revocation while Clarence was still alive (an invention Olivier concocted so that their silence in this later scene would attest to Richard's power over them; in Shakespeare's play Richard blames the late countermand on "some tardy cripple"). The King dies with the final blow of the news of Clarence's death (in the text occurring offstage between Act 2, scene 1 and Act 2, scene 2), and, after all exit, Gloucester and Buckingham seal their alliance. Standing alone in front of the dead King, they make a pact to join the retinue to Ludlow so that they may "part the Queen's proud kindred from the Prince." Richard *names* his cousin "my other self, my counsel's consistory," and Olivier reinforces the pact by shooting the shadows of the two men as they exit the room. Buckingham becomes shadow with Richard at the moment the alliance is named, and, in becoming shadow, he becomes an actor in Richard's drama. The isolated image of shadows in the filmic space of the frame underscores the very theatricality of this world.

Olivier's treatment of the alliance of Buckingham and Richard is significant here for another reason. The film suggests that while the former is a
conspirator, he is ultimately dependent on his more powerful cousin. Buckingham uses language and stages scenes with some independence but is
ultimately under the control of his great master. He merely adopts Richard's
ways for Richard's ends. It is interesting to note that while the alliance
between these two men exists, Richard does not confide in the audience,
does not look directly into the camera to take us into his confidence. He
ignores us. He seems now to have no use for us. It is only after King
Richard rejects Buckingham that he resumes speaking directly to his audience. Olivier points out how Buckingham's relationship to Richard is similar to the audience's relationship to the vice-figure. At first, Buckingham
delights in Richard's power and becomes his confidant. In Buckingham's
response to Richard we recognize our own lust for power, our own delight
in his compelling manipulations, our own desire to aid him in his theatrical
enterprise, and our own pleasure in the secrets he tells us. Buckingham is a
mirror of that very process. When Richard finally rejects him, he recognizes
that he made a choice to be deceived, just as the audience itself chooses to be
duped into participation with the vice. Once crowned, Richard no longer
has any use for his fellow conspirators and, in true Machiavellian fashion,
rejects those who helped him to power, lest he, like Henry IV, fall prey to
them in turn.

Olivier's treatment of the scene at Baynard Castle epitomizes how Richard uses and rejects his henchman and dupe. He opens the scene inside the
castle, where a sumptuous meal awaits the two conspirators. At the start,
Richard appears like a nervous actor waiting backstage for the moment of
his most important performance. As the scene unfolds, this room in
Baynard Castle serve as a kind of "green room." He anticipates the arrival of
Buckingham to hear of the Duke's encounter with the citizens after spread-
ing the rumor of the Prince's illegitimacy. When Buckingham finally enters,
Richard eagerly asks "How now, how now, what say the citizens?" Buckingham walks around the room, gathers food; and, like Juliet prying the Nurse
for information about the arrangements with Romeo, Richard follows his
cousin around the room with the impatience of an excited child. Finally,
immediately after Buckingham gives his news, Richard leaves to stand
above with a prayer book between two churchmen, while his conspirator
walks outside to manipulate the crowd from below. Olivier's treatment of
the rest of the scene, with its studio look and stage imagery, represents the
culmination of Richard's theatrical rise to power. The filmmaker chooses
not to fill the space with masses of citizens to create a realistic feeling for the
scene but purposely limits numbers to keep the atmosphere theatrical, using
no more than, say, twenty-five, which, in the context of the scene, looks like

Laurence Olivier's *Richard III*. Richard, next to clergy, looks down upon Buckingham and the citizens in the scene at Baynard Castle.

a stage crowd. As in the film of *Henry V*, *Richard III* maintains a sense of the
very artificiality of the theater itself to maintain the self-referential quality
of the original text. After Richard accepts the "offer" and the citizens dis-
perse, one of the churchmen begins to ring the bells of Baynard Castle.
Richard hastily pushes him aside, grabs the rope, and slides down to Buckingham and Catesby below. We watch in close-up the bells ringing and
turning over chaotically (an ambiguous image of celebration and upheaval),
and in the space of the tight shot we find ourselves conscious of being in the
realm of film once again. Richard's first and most important act as "king-
elect" then follows; he must assert his will over those who have placed him
on the throne. Buckingham approaches Richard to congratulate him. Richard offers his black, clawlike hand for Buckingham to kiss. But as Buckingham approaches, Richard gestures to the ground and forces the conspirator
to his knees. The other remaining nobles follow suit. Richard has the
power. Those who assisted him will enjoy no special benefits. In a theatrical
tableau, all kneel in obedience.

To simplify the formula Olivier adopts in *Richard III* would be to isolate
close-ups, quick cuts, high- and low-angle shots (which he uses especially in
the scene immediately after Richard is crowned), special effects (the translu-
cent ghosts of Act 5, scene 3); and to contrast them with theatrical set
pieces, long shots, stage tableaux, windows, doors, and archways, all of
which give the *mise-en-scène* a theatrical feel. The result of his approach is to
bring attention, through defamiliarization, to Richard's theatricality, which,
in and of itself, is Shakespeare's own device for creating a theatrical self-

consciousness in the play. Ironically, the filmmaker emphasizes the theatricality of his film through the new spatial relationships he builds, relationships possible only in the cinema. He calls attention to the device of his medium in a way that parallels Shakespeare's own attempt to do so with the stage.

III

The Chorus's call for imaginative participation to create the space of the *Henry V* epic is the center of Shakespeare's theatrical self-consciousness in that play; Richard's theatricality as one who writes the script and controls the space of England's history is the center of similar reflections in *Richard III;* Olivier informs both films with a dynamic of filmic and theatrical space to bring out those issues. *Hamlet*, on the other hand, is, relatively speaking, the most "cinematic" of Olivier's Shakespeare films. He repeatedly uses devices unique to the medium. *Hamlet* is a film of close-ups, extensive camera movement, flashbacks, and multiple angles and perspectives, all combined with a dominant soundtrack. To announce the arrival of the Ghost, for example, Olivier uses the sound of a heartbeat and dramatic music combined with shots (of Hamlet's terrified face) going in and out of focus in a pulsating rhythm. In presenting the soliloquies, he uses voice-over and close-up as the commonest techniques. Whereas the set for *Richard III* is largely flat (when the director does use levels, he uses them in a way that recalls the theater), *Hamlet* takes place in a world of curving stairs, multiple rooms, cliffs, and ramparts that we often see through the filmmaker's traveling camera. Yet a sense of the "staginess" of Olivier's *Hamlet* is undeniable, and critics traditionally focus their comments on that very aspect of Olivier's treatment, some with reservations.[9] The filmmaker juxtaposes moments in which the set appears wholly filmic with others in which it strongly resembles a stage set. He contrasts a restless, moving camera with moments of stasis that expose "blocked" scenes and precisely choreographed stage tableaux. His lighting varies in style to suggest both theatrical (spotlights) and filmic worlds. Bernice Kliman points out in her comparison of the filmic and theatrical characteristics of the film that, ultimately, "Olivier has created a hybrid form, not a filmed play, not precisely a *film* but a *film-infused play*, a form he conceived for being the best possible for presenting the heightened language of Shakespeare."[10] How does the specific juxtaposition of theatrical and filmic spaces call attention to Shakespeare's meditation on his medium and Olivier's reflections on his?

Shakespeare's exploration of the art of the stage finds its greatest focus in the first scene with the Players and in the "Mousetrap" episode that follows. In the Hecuba speech Hamlet marvels at the ability of the actor to engage so

Laurence Olivier's *Hamlet*. Laurence Olivier as Hamlet, Basil Sydney as Claudius, Eileen Herlie as Gertrude, and Jean Simmons as Ophelia.

passionately in the illusory, to become so transfixed, "and all for nothing!" Given his own "cue for passion," he is driven to chastise himself by comparison. But the theater's capacity to stir emotion is not only a capacity of the actor, it is equally a stimulus for the spectator; and so Hamlet devises his plan.

> Hum, I have heard
> That guilty creatures sitting at a play
> Have by the very cunning of the scene
> Been struck so to the soul that presently
> They have proclaim'd their malefactions;
> For murder, though it have no tongue, will speak
> With most miraculous organ. I'll have these players
> Play something like the murder of my father
> Before mine uncle. I'll observe his looks;
> I'll tent him to the quick. If 'a do blench,
> I know my course.
>
> (2.2.589–599)

What Hamlet expresses at this moment is part of the same paradox evident in his wonder at the Player's ability to weep for Hecuba, except that now he speaks of the spectator; the theater, despite its illusory nature (or perhaps because of it), has the power to stir the individual in a unique way. The stage event thus has a dual capacity: it is at once without substance and indicative of the most essential matter; a process that, despite its "nothing-ness," can speak for what has "no tongue"; an appearance that has the capacity to expose reality. In *Hamlet*, it is "The Mousetrap"—the play, the illusion—that ultimately unmasks the truth of Denmark and of Claudius's deed, the device that confirms the Ghost's story. For Shakespeare, the stage is obviously not an imitative reality, but a place of self-conscious illusion that can stir the passions of spectator and actor alike.

The moment in the film, therefore, when Olivier most clearly works with a dynamic of filmic and theatrical space to explore Shakespeare's own reflec-tions on his medium is, appropriately, in the "Mousetrap" scene. Onstage, the play-within-the-play is interesting on many levels, not the least of which is the way it comments on the play proper. "The Mousetrap" is perfor-mance criticism in motion, commenting through the process of a perfor-mance on the nature of theatrical activity—all the while exposing a deed of the past and the depravity of the *Hamlet* world. It is an event in this tragedy similar to the moment when the Chorus of *Henry V* calls on the forces of imagination: both are devices the playwright uses to reflect on his art.

Curiously, Olivier cuts, almost entirely, Act 2, scene 2, in which Hamlet greets the players, listens to the speech on Pyrrhus, asks the company to perform the "Murder of Gonzago," and chastises himself in the "Hecuba" soliloquy. In the film, Polonius finds Hamlet sitting alone in a dark room and announces the players. The Prince responds abruptly, "He that plays the king is welcome" (he must already have concocted the "Mousetrap" plan), greets the players in a brief interchange, and asks the leading player to put on the "Murder of Gonzago" with the insertion of "some dozen or sixteen lines." The players quickly exit and Hamlet runs to a small platform in the room, stands in a spotlight, strikes an unmistakably theatrical pose (in a long shot), and calls out in a fully projected stage voice: "The play's the thing / Wherein I'll catch the conscience of the king." The filmmaker obliterates the details of Shakespeare's meditations (through his poetry) on the nature of theatrical performance and its peculiarly penetrating qualities. Instead, Olivier relies on the events of the "Mousetrap" scene itself, unfolding in a dynamic of filmic and stage space, to articulate the very issues that Shakespeare raises in the scene generally and in the "Hecuba" speech specifically.

Hamlet's speech to the players takes place in a space distinctly filmic. We hear his opening words ("Speak the speech, I pray you") in voice-over while the camera pans the players listening earnestly to his advice. The camera finally finds Hamlet on "O, it offends me to the soul," but then quickly leaves him to focus on the props, masks, and instruments of the company while the words continue in voice-over. The isolation of aural and visual fields, the close-ups, the panning camera—all contribute to making this an event that unfolds in a filmic realm. But with the entrance of the royal train,

Olivier shifts to an atmosphere of theatricality. In a long shot that exposes an enormous area, we watch, to the sound of a flourish, the entrance of the King, Queen, courtiers, and others, in a procession down the dramatic curving stairs that lead into the room. Indeed, the moment most closely resembles, with its stationary camera exposing the whole picture, a stage entrance. When the play-within-the-play begins, the camera roams freely, following a semicircle behind the arrangement of spectators, first shooting from behind Horatio's shoulder, then moving to the center to shoot from behind the King and Queen (catching the silhouette of Claudius giving Gertrude a quick kiss), then moving to the perspective behind Hamlet and Ophelia, exposing all the while both the play and the shadows of the figures watching the play.

Already the layers of filmic and theatrical spaces form a compelling dynamic. The camera moves with the freedom of cinematic space as it shoots "The Mousetrap," just as Olivier's camera can roam with similar freedom of movement as it shoots Shakespeare's *Hamlet*. In addition, just as the camera on "The Mousetrap" (a stage enactment on screen) incorporates both filmic and theatrical attributes, so, too, does Olivier's cinematic treatment of Shakespeare's tragedy. The frame of the pictures exposed by the moving camera, however, also exposes the silhouettes of the spectators watching the performance of "The Mousetrap"; similarly, the spectators of the cinema proper (unless they are sitting in the front row) also have in their vision the silhouettes of their fellow audience members in front of them watching Olivier's *Hamlet* (not to mention the fact that if they are not in the last row they are also silhouetted figures in the view of others behind them). The film viewer finds himself or herself in a chain of observers, watching (and being watched by) others in the cinema in the same process of watching the Olivier film depicting a scene of characters in *Hamlet* watching each other watch "The Mousetrap." The levels of observers in different realms and the film's ability to call attention to the layers of filmic and theatrical activity show how Olivier can use his medium to parallel Shakespeare's own musings on the processes of observing a theatrical event and the peculiar nature of the art itself.

As the scene continues, Olivier juxtaposes filmic and theatrical space with alternating shots of the play-within-the-play and reaction shots in close-up of Claudius, Gertrude, Hamlet, and Horatio. Moreover, the filmmaker deliberately chooses to show only the mimed prologue of "The Mousetrap" (Claudius bursts out in anger and panic before they can get to the dialogue) and this decision also contributes to the special mixture of filmic and theatrical spaces. The prologue without words uses images and pictures in a theatrical context and thus mirrors the structure of the film itself at this moment. Olivier's wordless images on the screen combine with the wordless prologue of the play to tell the story.

Laurence Olivier's *Hamlet*. The silhouettes of Claudius and Gertrude as the camera roams behind the spectators of the play-within-the-play. The film viewer is part of a chain of observers.

As the "Mousetrap" sequence began in a theatrical space, with the long shot of the royal entrance, it ends in a filmic one. We see Claudius explode after the camera circles him and scrutinizes his reaction. The King rises. Through a low-angle close-up on his face, we see him put his hands over his eyes in panic and then scream, "Give me some light." The following cut is to a full-frame shot of a torch moving across the room—but the camera eventually tilts to show Hamlet to be the bearer. In a close-up shot, Hamlet thrusts the flame in Claudius's face (literally exposing him with light as the play has done metaphorically), forcing him to flee. A sequence of shots in rapid succession, of the King's departure and of members of the assembly running off in multiple directions, follows to visually articulate the chaos that ensues. We enter fully into the realm of cinematic space.

One other point about the "Mousetrap" scene is pertinent here. In the film, this is not the first time the spectator witnesses the image of Claudius poisoning King Hamlet. In the first act, Olivier visualizes the event as a flashback while the Ghost, in voice-over, tells Hamlet its story. Significantly, the filmmaker blocks the murder in both the flashback and the play-within-the-play in an almost identical way. The significant difference is that, in the flashback, Olivier deliberately avoids showing the face of the murderer, avoids revealing his identity, and thus maintains some ambiguity about the truth of the Ghost's story. In other words, Olivier avoids making the test of "The Mousetrap" redundant in the film. More important, however, is the fact that the

information communicated in the filmic space of the flashback finds completion in the theatrical space of "The Mousetrap," just as the Ghost's words in
the place achieve verification in the play-within-the-play.

The repetition is significant for another reason. As the Ghost begins its
story, Olivier zooms in with a shot of Hamlet's head from behind, suggesting that we are entering the hero's imagination as he visualizes what the
Ghost recites. Because the filmmaker presents the flashback through the
filter of this shot (of Hamlet's head), he unambiguously asserts that what the
spectator sees is Hamlet's image of the murder, a subjective visualization of
it and not necessarily the true scenario. In the film as much as the play, we
cannot know for certain (not until "The Mousetrap" and the prayer scene)
that the story of the Ghost is authentic (even though Claudius expresses the
burden of his guilt in an aside in Act 3, scene 1, lines 50 ff). Logically,
therefore, the "Mousetrap" scenario is a stage enactment of Hamlet's subjective picture of the event. Granted, the Ghost is quite detailed in his description and one could not imagine a visualization that differed markedly from
the one Hamlet imagines in Olivier's film. Still, the important point is the
connection between the two moments and the implication that "The Mousetrap" is a product of Hamlet's visual imagination. (The film, in fact, inadver-
tently points out that this is implicit in Shakespeare's text as well.) Olivier
makes his point through the specific juxtaposition of filmic space (and its
accompanying internal subjectivity) with theatrical space (and its outward
and public realization of that internal picture). Shakespeare demonstrates
how the theater, as a product of the imagination, can unmask the reality of
Elsinore; Olivier demonstrates how the resources of the film medium can
expose the very processes at work and formulate connections that allow us
to probe more deeply into the playwright's meditations on his art.

IV

Peter Brook discusses with fascination the key moment of Gloucester's
"suicide" in *King Lear.* He sees it as an episode that crystallizes the infinite
potential of Shakespeare's stage and, by implication, the problem with
adapting the work to the screen.

> Kott's great essay on Gloucester's suicide is marvelous from that point of view.
> He points out that the suicide only means something if he does it on a bare
> stage without a rock to jump from, because then it becomes the whole of
> Pirandello and the whole of Ionesco and the whole of Beckett. It is a man doing
> a meaningless jump, and an actor doing a jump, at one and the same time. . . .
> If you take the scene of Gloucester's suicide you are forced, in the theatre, to
> make Gloucester do it on a windy heath of some description. Fifty per cent of
> the extraordinariness is that it is happening on an imaginary heath and yet on
> the boards of the theatre. There's a meaning there that is released by that

double tension which isn't there if you take either aspect on its own. If it's just a leap on a bare stage, it hasn't the meaning; and if it is really a man on a heath doing a leap it also hasn't; but in Shakespeare, without any effort at all, you get both. It's like an idea itself striking you.[11]

The "double tension" that Brook discusses is, in a very important sense, a product of the theatrical self-consciousness that informs Shakespeare's work. The moment operates on the basis of the interplay of description and imaginative building: in his description of the cliffs, Edgar does for Gloucester what Shakespeare and his actors do for the audience at the Globe. In both cases, flat worlds are magically transformed in the dance of language and imaginative fancy. The self-referential theatricality of the moment is part of its power; and Brook asks, seven years before he made his film of *King Lear*, how it would ever be possible to give meaning to the moment on film.

When he does film the scene, Brook's ultimate strategy is not unlike the one Olivier adopts to bring out a filmic self-consciousness in his work: Brook creates a dynamic of theatrical and filmic spaces that operates in relationship to Shakespeare's language. "The director duplicates in cinematic terms Shakespeare's blend of blatant stage artifice and imaginative reality."[12] The sequence begins with a blurred shot of two figures, in silhouette, approaching the camera. To us the world is out of focus and lacks detail, a technique that allows us to share, through filmic convention, Gloucester's blindness. As Gloucester begins to speak, Brook shoots a white sky with the sun only intermittently visible behind rapidly moving clouds, an image of occluded light as well as one that suggests a visual and linguistic pun on the sun/son to which/whom Gloucester is blind. The clouds then thin slightly, seem to merge with the white light of the sun, and the entire picture swells in bright, "blinding" light (in opposition to the black of the storm scenes). We hear on the soundtrack Gloucester's request: "There is a cliff, whose high and bending head / Looks fearfully in the confined deep. / Bring me but to the very brim of it, . . . " The movement in the tones of light of the sky throughout this sequence mirrors Gloucester's experience; it suggests both blindness and a light of guidance he is not aware of. That Edgar is indeed associated with the bright light is clear, for his face in full close-up eventually enters the white, nondescript frame to specify what visual metaphor could only suggest. It is no accident, moreover, that Brook centers Edgar's face in the frame in exactly the place where we saw the "sun" a few frames earlier; the pun is complete.

Thus far, then, Brook has placed the action of the sequence in· filmic space. He continues this through the next shot: we see, in close-up, the feet of Edgar and Gloucester running and "struggling" on the ground, the "hill" that Edgar describes. Here Brook's compelling technique is already evident:

Peter Brook's *King Lear*. Brook's shot of Edgar holding Gloucester demonstrates how the director uses "filmic" space to create the illusion of the men on a cliff. The spectator has no point of reference to measure the flatness of the land they walk on.

the filmmaker plays with the visuals so that the spectator has no point of reference to measure the land the two men walk on. All we see are the moving, tired, lower legs in a tight shot. We hear, however, Edgar's description. Thus, whatever slope we may conjure in our minds with Gloucester— or whatever tension may arise through imagining that Gloucester, and the actor playing the role, creates an imaginary hill, through a peculiar necessity to believe—that slope is a product of language and the ambiguous visuals. Shakespeare increases the tension of the moment in the theater with the surprising reminder that comes from Gloucester's words, "Methinks the ground is even." They are words, ironically, that Gloucester speaks just before he commits himself to believe he is climbing; it is a moment of doubt, a question that demonstrates paradoxically that the imagination is ready to engage. As in the theater generally, the questioning, the very awareness of the flat and unencumbered world, is an essential element in the freedom we assume to complete the picture in our own imaginations. Without his expression of doubt as a step in the process, Gloucester's subsequent belief in reaching the top of Dover would never be so convincing to the spectator, nor would we be so cognizant of the levels of theater that inform the moment.

Brook builds a similar tension in the film with a cut to a shot of the men's backs (at the moment Gloucester speaks these words) and a reverse zoom to reveal them on a flat surface. With this picture, we are suddenly in theatrical space, a space that defamiliarizes the filmic space of the previous shots and thus shocks us into an awareness of the medium operating. In the play, Shakespeare comments on the nature of the theater by creating a moment in the drama in which one is keenly aware of how words transform the space of a flat surface; Brook follows Shakespeare with a comment on the nature of film by creating a moment in which he reminds his spectator, by the long shot, of the flat surface the men walk on. Or, to put it in the terms of this chapter, he brings out the transformations in filmic space by defamiliarizing that space with a theatrical contrast.

We return to filmic space in the following low-angle close-up shot of Gloucester riding on Edgar piggyback. Again, the technique Brook employs to create the illusion of climbing is one that does not allow us to detect the ground or the horizon. With Gloucester, we cannot see the ground, and we build its topography on the basis of what Edgar describes in combination with the open-ended visuals Brook offers. In addition, within the close field, we share an intimate space with these men and perceive in Edgar's face the pain of the son who enacts this ritual for his father. Just as Gloucester is about to fall, Brook places us firmly in cinematic space with a full close-up on the despairing man. The ultimate point is made in the next shot of Gloucester, falling on the flat ground, an aerial shot that alienates us from the close perspective of the filmic world and exposes a flat, theatrical space. Brook builds his own "double tension" through the dynamic of spaces he offers. "It's like an idea itself striking you."

Like Olivier, Brook recognizes the centrality of Shakespeare's self-conscious musings on the nature of his art as an integral part of his dramaturgical design. What the playwright achieves through words and action, the filmmaker imitates with a juxtaposition of spaces that comment on each other. Ultimately we are aware, in both instances, of the nature of the medium before us, and of the necessity of our own creative participation with it. By defamiliarizing their respective arts, playwright and filmmaker encourage and enrich the imaginative activities of the spectator.

CHAPTER 6

Temporal Multiplicity: Patterns of Viewing in Cinematic Time

I

Just as cinema offers the play a new spatial field for its realization, so, too, does it give to the drama a specific temporal organization. The question of the plays as a product of filmic time, however, is a subtler and slightly more difficult one to address than that of space, partly because Shakespeare's own temporal structure is so close to that of the film medium itself. Indeed, one of the first and most striking ideas to emerge from early studies of Shakespeare on film was the apparent similarity between the form and structure of Elizabethan plays and the basic elements of the cinema. The fluid nature of time and space in Shakespearean drama suggests remarkable links to the non-linear, open structure of the motion picture; as with film, laws of cause and effect and verisimilitude do not control the structure of Shakespeare's plays. One critic, Henri Lemaitre, went so far as to discuss Shakespeare as a "cinematic writer" in a "pre-cinematic age."[1] Shakespeare moves from place to place with a freedom limited only by his pen and imaginative fancies. Similarly, he creates a temporal framework free from the strictures of everyday perception, one that ranges from the tight and relatively "unified" events of *Othello* to the sixteen years of *The Winter's Tale*. But whereas cinema creates spatial fields through multiple perspectives, context, the range and distance of shots, the moving camera, dialectical relationships between *kinds* of space, and so forth (all of which give a specific spatial organization to the already "open" structure of Shakespeare's plays), the way a filmmaker works with temporal strategies is not nearly so apparent. The spectator's imaginative process to accommodate the evolution of time in the progression of the events in a film version of, say, *Hamlet* or *King Lear*, is not very different from the processes involved with stage production or reading. Unlike the manner with which they develop qualities of space that are distinctly cinematic, directors of Shakespeare films rarely accentuate the equivalents of time.[2] To date, there are relatively few flashbacks in Shakespeare films, few instances of the expansion or reduction of time when not already part of Shakespeare's script, few examples when the duration of a scene or event is fundamentally different from its implied duration in the text (unless the scene is rewritten specifically for the film), little use of slow or fast motion, lapsed time, or high-speed photography. In short, it is difficult to isolate for the temporal field what a technique like the close-up does to the spatial organization of the drama.

How, then, can we begin to discuss the way the spectator works with the elements of the Shakespeare films to build a temporal structure? How is that process somehow unique, or, more important, how do the temporal strategies of the filmmakers activate a peculiar imaginative dynamic? To address these questions I will borrow again from Iser's phenomenological strategy to clarify a paradigm common to film and stage (and reading), consider how it operates in Welles's *Chimes at Midnight*, and then proceed to look at how we might begin to understand its purpose in a cinematic context.

Iser defines the temporal process of reading as an activity of anticipation and retrospection. He examines how sentences act on each other in the reading process, how they are both informational statements and expressions of something that is to come, an indication of a future. He argues that sentences in combination work to "shade in" what is imminent and produce expectations that, in a non-didactic work, the text continually modifies. This notion of continual modification leads to Iser's definition of a process of anticipation and retrospection. We not only move forward with the sentence in "anticipation", but move backward as the statement in the present causes to revise in a process of "retrospection":

> each intentional sentence correlative opens up a particular horizon, which is modified, if not completely changed, by succeeding sentences. While these expectations arouse interest in what is to come, the subsequent modification of them will also have a retrospective effect on what has already been read. This may now take on a different significance from that which it had at the moment of reading.[3]

Iser concludes, therefore, that there is always a dynamic operating for the reader between present perception and memory, which in turn activates a continual modification of future expectations.

Iser's model translates well for the process of viewing a performance on stage or screen; surely, the performance "text" activates a similar means of building expectations for a future that are continually modified by memories of what is passing and has passed. The primary unit for Iser is the sentence (or, as he labels it, "the intentional sentence correlative"), but the experience of a performance moves far too rapidly to isolate such a small unit of expression. The reader of a book moves at a pace he or she controls and can dwell on an individual sentence or word (or move quickly past it) by choice. In contrast, the viewer of a film does not control the speed at which the units of performance evolve. To isolate a basic unit of operation in a performance is perhaps impossible, because a performance is always "on the move," beyond the control of the audience.[4] Still, the viewer is nonetheless involved in a temporal odyssey as fluid as the one of Iser's reader. What interests me specifically about this process is the multiple levels of interruptions that make up the temporal framework of the film; in the imagination, each viewing moment is constituted by memories of the past and expectations for the future that shake the seeming solidity of a temporal present. Indeed, it is difficult, if not impossible, to define a unit in time without recognizing its fundamentally liquid and unsteady nature. Temporal interruption is a necessary part of viewing performance, and in Shakespeare films it leads specifically to curious forms of simultaneous action, thematic association, visual substitutions, and aural and visual dynamics.

Orson Welles's *Chimes at Midnight* is fertile ground to begin an exploration of this notion of temporal interruption, partly because of its scope (it is a film that operates as a broad story of one man's journey in time from saturnalian misrule to old age and death) and partly because Welles creates central moments in the film when meaning functions precisely through the viewer's process of anticipation and retrospection. In fact, as I indicated in a previous chapter, one strength of this film lies in its ability to sustain a primary tension for the viewer between a sympathy for Falstaff and a nagging and ineluctable sense that he must be rejected. Welles develops and sustains this tension through the temporal field of his film, whereby memory of one scene mixes with the experience of viewing in the moment, a mix that accounts for the modification of our expectations of what is to come.

It is because Welles develops such a precise temporal strategy that, intentionally or not, he checks the film's (and his own) tendency to sentimentalize Falstaff, whom he has called the "good, pure man."[5] Welles has confessed to a rather profound identification with the Knight and, as Jorgens points out, it is not difficult to understand why Falstaff's story might chime with Welles's own tragic self-image.[6] One must caution, however, against basing one's analysis of the film on Welles's words rather than on *Chimes* itself, a mistake many critics make.[7] Although Welles elicits our sympathies for Falstaff, we do not see him as "a good, pure man," but as the witty, colorful, part braggart, part vice, and part father-figure, the entertaining and self-centered "trunk of humors," "father ruffian," and "vanity in years" that he is. Indeed, Welles portrays the many sides of the Knight by deliberately creating moments that work in a dynamic through the journey of viewing.

As I mentioned earlier, Iser defines a didactic piece of writing as one that confirms expectations rather than modifying them. A melodramatic film of the Falstaff story would, in a manner similar to that of Iser's didactic text, take us to the rejection scene without stimulating any ambivalence about Sir John's plight. We would sympathize with the central figure and never question what we would assume to be his cruel and undeserved misfortune (as many romantic critics, especially since Maurice Morgann, argue about the Falstaff story).[8] Welles's film elicits no such thing. The filmmaker evokes a complex and ambivalent response throughout. The relationships among the critical moments Welles creates, relationships that are formed and continually modified through the viewing process, are at the center of the spectator's sustained ambivalence. In other words, the temporal dynamics of the film demystify the very sentimentality associated with the story of the fat Knight, sentiment that Welles himself seems to promote. Let us look at a specific example.

In his article on *Chimes at Midnight*, Samuel Crowl focuses on Falstaff's relationship with Hal and isolates central scenes (alluded to in an interview with Welles)[9] in which Hal bids his acquaintance farewell.[10] Crowl demonstrates how Hal's first soliloquy ("I know you all . . . "), the Gadshill episode, the scene including the "play extempore" ("I do, I will") and the Prince's departure from the tavern, the end of the Knight's speech celebrating sherris-sack—all combine with the final rejection scene itself to build one "long goodbye." Crowl's sensitivity to the filmmaker's strategy of creating moments that work through association helps to illuminate the imaginative process that takes us through the film as a whole. He demonstrates how the director stimulates the viewer to work with individual parts in the activity of realizing the filmic text.

The moments of farewell function as a dynamic through time. The spectator views, say, the "play extempore" as a moment that anticipates the final rejection and yet, simultaneously, as a moment that evokes the memory of Hal's deliberations in his earlier soliloquy: "Redeeming time when men think least I will." The evocation of that memory then colors the present experience of the play-within-the-play, our expectations for what is to come in the final rejection itself, and, retrospectively, our experience of the soliloquy. Throw into the dynamic other moments of farewell, and the imaginative dance of past, present, and future creates the temporal multiplicity of the film. It is important to note that in these farewell scenes, Welles triggers temporal connections through both form and content; in addition to the content of the "goodbye" theme, a precise repetition of the shot/counter-shot pattern with the close-up on the faces of Falstaff and Hal activates this imaginative activity. At the moment of "I do, I will," Welles isolates Falstaff's face for several beats: "the camera catches . . . the same quizzical expression that we witnessed peering at us in the background of the 'I know you all' soliloquy."[11] Imaginatively, memories and expectations interrupt each moment of farewell, and, like the multiple angles of the spatial field, the film creates a corresponding tapestry in time.

Through the kaleidoscope of perspectives, expectations, and recollections, Welles prepares us to feel both the pain of separation and the need for Hal to restore what we have come to recognize as a diseased nation, made most evident through the visual brutality of the Shrewsbury battle. (As I pointed out in a previous chapter, the dynamic of spatial perspectives in the rejection scene also participates in activating our ambivalence.) Thus the final rejection, in addition to functioning with scenes of farewell, also functions with the specter of other moments that give rise to both sympathy and antipathy for both men. In retrospect we recall the lovable Falstaff of the Boar's Head as well as the coward in war; we remember the Knight's recruitment of his lowly soldiers and his acceptance of bribes, which, in the context of the battle as Welles presents it, accentuates the unmitigated exploitation of a "lower" social class; we recall the moments of tenderness that Falstaff shows to certain figures (notably Hal and Doll Tearsheet), his charming self-criticism and honesty about his frailties, but also his repellent self-centeredness and greed. Similarly, we delight in Hal's wit and his sense of play—both made vivid through the film—and long to see him in Falstaff's company. But we also remember the atmosphere of disease and malaise of this world, which Welles accentuates in the latter half of the film, as well as the tyranny of civil strife; in this production, Hal is the only hope. Thus, the final rejection, a dramatic climax of the film, operates through a tension of sorrow and hope, a tension that is the result of the spectator's imaginative work with the temporal field of the film.

II

While Welles repeats key visual patterns to realize the temporal multiplicity of *Chimes*, the fundamental paradigm of anticipation and retrospection holds for any art form that moves "through time." The cinema can work with a play in a more specialized way, not so much through diachronic relationships, but through synchronous ones. Film can fragment a moment of Shakespearean drama to create a specific dynamic; like cinematic space, conventions of cinematic time can affect the experience of the play by creating many perspectives for a single instant, an experience of simultaneous action. We have thus far seen how temporal interruption operates in the specific imaginative realm of memory and expectation through the entire scope of the viewing process. But film can also divide the smallest moment by interrupting the main narrative line with a convention that suggests events occurring simultaneously. Paradoxically, the juxtaposition of simultaneous events, though formally linear in film (images unfold in a linear sequence), displaces temporal linearity by eclipsing normal cause-and-effect relationships. Synchronous action activates the imagination in a way that defies temporal progression in everyday experience by depicting, in a linear progression, a single point in time. Macbeth speaks of an imaginary dagger; Welles intersplices shots of the witches flashing a dagger before the eyes of the figurine. We "read" the image of the witches here to be an event simultaneous with Macbeth's hallucination that offers a comment on the major action and engenders a process of anticipation and retrospection within the moment. Multiple perspective now takes on a distinctly temporal flavor.

Curiously, the live stage, and not the screen, is the perfect habitat for simultaneous action because the broad perspective of the theater allows for two or more activities to appear on the stage in different places but at the same time (as Shakespeare himself demonstrates in the last act of *Richard III*). In film, however, we must rely on conventions of simultaneity as the images appear in sequence (with the exception, of course, of a split screen, which is not used in the Shakespeare films under discussion). The sequences that suggest concurrent activity produce interesting results specific to film viewing. In the following example from Kozintsev's *Hamlet*, one recognizes how the filmmaker develops such a temporal framework to make a clear political point about Claudius's deeds.

The opening of Shakespeare's *Hamlet* is a time of rashness and anxiety, a time in which the Queen has joined with the brother of the dead King in an "o'er hasty marriage." In the play, Marcellus tells us that the constant bustling of activity in Denmark "Does not divide the Sunday from the week." He asks "What might be toward, that this sweaty haste / Doth make the night joint-laborer with the day?" The officer's words call attention to the destruction of the rhythm of the week, to a world where the activities of night and day find no contrast. Although he cuts Act 1, scene 1 from his film, Kozintsev builds a sequence of images that duplicates Shakespeare's expression of haste and multiple activities through a prologue made up of cannon fire, Hamlet galloping to Elsinore, and servants hanging black banners of mourning from every window of the castle. All of this activity is set against a backdrop of shadows, ominous clouds, and the shivering choral tones of Shostakovitch's score. The whole comes across as a strange mixture of efficiency and anxiety, enacting the proper and familiar ritual in a spirit of terror and dread. The spectator witnesses a series of strange events in a rapid sequence of images and senses the tension of a broken rhythm, yet fails to understand the cause. Then, with Claudius's speech in Act 1, scene 2, we begin to get some information.

Kozintsev's treatment of this opening speech is fascinating for the way it unfolds as a function of simultaneous action. Instead of observing Claudius speaking his opening lines in court ("Though yet of Hamlet our dear brother's death"), we hear the words on the soundtrack while the camera pans crowds of peasants standing outside the castle walls. Eventually, Kozintsev's camera focuses on a soldier reading the words to the peasants as a royal proclamation. The world listens attentively to the statement of the King—a world of people who, like the audience itself, crave answers to the mysterious events that have passed. On the line "In equal scale weighing delight and dole—/ Taken to wife," Kozintsev cuts to courtiers inside Elsinore repeating the phrase "in equal scale" on their way to the assembly. The courtiers overlap the soldier's words, suggesting simultaneous activity inside and outside. As the courtiers exit, two German ambassadors enter the frame, also repeating (in a tone of inquiry) the phrase in German, followed by two French ambassadors quoting Claudius in French. Finally, we are in the assembly, where Claudius himself continues the speech, beginning with "Nor have we herein barr'd / Your better wisdoms, which have freely gone / With this affair along." In context, and by introducing Claudius's speaking already in progress, Kozintsev establishes that the King speaks simultaneously with those outside the immediate court. The scene remains in the assembly until Claudius turns his attention to Hamlet.

Kozintsev chooses to highlight the political implications of Claudius's speech by placing it in a unique temporal context, by creating the illusion of simultaneous action that makes the speech an event of multiplicity in a single moment in time. The court—and all of Denmark—is buzzing with the "statement" of the new king, and Kozintsev's shots give the spectator a sense of concurrent activity. The spectator in turn works with the elements of the moment in time through the modifying process of anticipation and retrospection, a process that gives this speech of Shakespeare's text a complex fabric of temporality. The repetitions of the courtiers and ambassadors work with the memory of the soldier outside, which itself anticipates the tyrannical figure that is Claudius in the film. The sequence exposes the multiple dimensions of a single moment, the moment when a tyrant speaks. Kozintsev relocates Shakespeare's words in time and space, pulls them out of the context in which we witness them on stage, and by doing so gives the King's declarations a special resonance: "never alone / Did the king sigh, but with a general groan." Claudius's deeds affect everyone.

The implication of simultaneous action is equally significant later, at the moment when Hamlet, Horatio, and the soldiers search for the Ghost on the dark battlements. In the midst of their wait, Kozintsev abruptly cuts to the loud, bright, boisterous celebration of the marriage festivities below and inside. What Shakespeare treats as an offstage activity, Kozintsev foregrounds in a fully realized scene to create a simultaneity of dramatic action. The music, dancing, laughter, and raucous energy collide with the ominous figures on the clock,[12] the dark of the night on the battlements, and the search for the spirit of the king of the past. The director shoots Claudius pulling his new bride into a private room to consummate their union. Then he cuts back to Hamlet on the windswept battlements waiting for the ghost of Gertrude's former husband. At the moment the Ghost arrives, therefore, the search for the past (the image of the dead King Hamlet) and the ineluctable force of the future (made most explicit through the image of the clock with the circular movement of figures followed by Death, not to mention the marriage feast inside) combine in present action, action that unfolds in the simultaneity of cinematic form. The moment becomes a function of cinematic time when past, present, and future commingle in synchronous events of celebration and dread; we perceive "the character of a feast in time of plague."[13]

What also emerges from these examples of simultaneity in Kozintsev's film is a curious process of substitution in which images that unfold in sequence "stand in" for and color each other. The filmmaker's initial presentation of the Ghost is a case in point. A series of shots of horses wild with fear in the stables accompany Horatio's terrified announcement of the Ghost: "Look, my lord, it comes." When we do see the Ghost, it is an enormous, black, towering giant, its height emphasized by an extreme low-angle shot. Hamlet, silent, follows as it beckons him. But Kozintsev fragments the scene with shots of the horses in a state of panic, breaking out of the stables. The shots of the terrified animals serve as a comment on the main action; they are the voice of the hero, the sound Hamlet is unable to utter at the moment he confronts the spirit who resembles his father. The spectator's experience of the moment when Hamlet confronts the Ghost is thus interrupted (in performance time) by these shots, and yet, by maintaining the illusion of simultaneous action, expresses a dimension of the main action. The film creates a "dynamization" of time in the imagination of the viewer as the images comment on and substitute for each other.

The sequence in Brook's *King Lear* depicting Goneril's growing discontent with her father and his hundred knights is another, more complex, example of how Shakespeare's dramaturgy finds expression by image substitution in the conventions of simultaneous action. From the first shot in Goneril's castle to the arrival of Lear from the hunt and the daughter's eventual confrontation with her father, Brook interrupts the narrative flow to trace the burning fuse of a bomb about to explode. The sequence unfolds through the following shots: (1) Albany and Goneril sitting alone by a hearth, a scene in which she expresses her frustration with her husband's "milky gentleness"; (2) a close-up of Lear and his men galloping recklessly at full speed; (3) a cut to a distant shot of this same scene that exposes vividly the size of the crowd that the King brings with him; (4) Goneril talking to Oswald inside about Lear's riotous knights and her order to Oswald that the other servants "neglect him"; (5) a long shot of Lear's train (we learn from the next shot, no. 6, that we are seeing the long line of knights from Goneril's perspective and hear her on the soundtrack add, with disdain, "idle old man"); (6) an aerial shot of the knights entering the castle; (7) another shot from Goneril's perspective of the unruly crowd in her house (she comments: "old fools are babes again"); (8) a short scene in which the disguised Kent offers his services to Lear; (9) a shot of Goneril announcing to all her servants, "let his knights have colder looks among you. / What grows of it, no matter"; (10) Lear's boisterous call for dinner and his fool.

The quick succession of these shots gives the sequence a specific texture of simultaneity as the depiction of Goneril's frustration and scheming alternates with the images of Lear's activities; in multiple shots, we witness the fragments of these two major lines of action. The key sequence of Shakespeare's play where the child plans to revolt against the parent functions in a continual dialectic of concurrent dramatic action. Every time we are with Lear, we bring with us a strong sense of the bubbling discontent inside; every time we listen to and observe Goneril, we realize that her rebellion is against a wild and charging energy. Brook prepares the audience to anticipate an explosion of the greatest magnitude. Indeed, simultaneous action sets in motion the conflicts to ensue, just as Kozintsev's traveling camera in the love contest gives the spectator similar clues in the spatial field.

In addition, the entire sequence develops at a gradually increasing speed. The very rhythm of Brook's montage adds, in a mode of "anticipation" *through* the film (as opposed to the single moment), to the sense of an incipient explosion. Interestingly, in the diachronic process of retrospection, the speed and chaos of the shots stands in direct opposition to the first third of a film we remember as often still and quiet. In the confrontations that develop in the first two acts, Brook deliberately contains the energy of his actors (with the exception of the riot in Goneril's castle as Lear departs); there are few outward explosions in this film as the unraveling of the family takes place. Brook contains the energy to build a pressure that is released only in the storm. The dual action of Lear's obstreperousness and Goneril's revolt sustains this tension and moves the play to the profound release of the storm scenes.

Kozintsev also creates simultaneous action in his *King Lear*, but his most powerful use of it occurs in a development of two lines of action he extrapolates from Shakespeare's text, a juxtaposition of events not quite so specific

in the original. Curiously, he avoids Shakespeare's own structure of simultaneity in the scenes of the storm and the blinding of Gloucester. The Soviet
director juxtaposes two major sequences, the stories of Lear and Gloucester,
rather than following the original to weave them in a tapestry of individual,
short scenes. Using associative montage he could easily parallel Shakespeare's technique of juxtaposing the storm with the blinding of Gloucester.
His decision to make the experience of Lear a single, uninterrupted whole,
however, allows him to contrast the totality of the storm outside with the
metaphorical one inside. Once inside Gloucester's castle, we witness a
storm caused, not by the elements of nature, but by the hard hearts of
human beings, and it is here that the filmmaker makes an important point
using simultaneous action.

In the blinding scene, Kozintsev creates a sense of chaos and turmoil
through a series of quick shots in montage, using cinematic form to realize
the tumult of the moment: the turbulence of the film's visuals continues the
storm indoors. Moreover, on the soundtrack, we hear a cacophony of
sounds: doors creaking and slamming, the heavy stamp of Cornwall's boots,
the grunts and groans as Gloucester is tied to a chair, all culminating in the
animal-like cry of the man blinded. Indeed, the inside noises are an echo of
the primitive sounds heard outside as the storm began. After a violent
interrogation of the old man, Cornwall pushes Gloucester's chair over onto
the floor. Rather than using the cinema to visualize the actual deed,
Kozintsev maintains distance and simply uses quick cuts to create a sense of
chaos. The camera momentarily shows the spike and heel of Cornwall, but
it pulls away at the moment the deed is done. In fact, as I shall indicate, the
technique of simultaneous action that Kozintsev employs in the following
shots fills out the horror of human terrorism far more than a bloody rendering could ever achieve.

Out of a terrifying moment of human bestiality, a servant miraculously
comes forward to defend Gloucester (can we anticipate some hope?). He
stabs Cornwall; in retaliation, Regan violently stabs the servant. The chaotic visuals continue as Gloucester's second eye is plucked out. His screams
are piercing and resonate through the castle hauntingly. The camera moves
through the halls of the castle as Gloucester's cries echo in emptiness. Then,
in an unexpected shot, we are taken to the location of concurrent action and
see Goneril standing next to an unmade bed, lacing her boot. She looks up
momentarily, hearing the cries of Gloucester, only to return to the seemingly more important activity of tending to her boots. On the soundtrack,
we then hear Gloucester cry, "Where's my son Edmund?" The following
shot shows Edmund, in his apartment, putting on his belt and dressing
himself. The implication is clear: while Gloucester's eyes were plucked out,
Edmund was upstairs with Goneril; evil reaches a point of consummation

not only in the union of these two figures, but in the more formal link of the two lines of action of terrorism and lust, a link, moreover, that functions in the film as a temporal equation. The juxtaposition of the cruelest abuse with the sexual act, of an act of vicious destruction with one associated with creation, of one man's torture and another's simultaneous pleasure, sums up superbly the twisted and demented universe of *Lear*.

As the blood runs down Gloucester's cheeks and he calls for Edmund, we remember his earlier statement to Kent regarding his bastard son: "I have so often blush'd to acknowledge him." Now the blood is external. The "good sport" of Edmund's "making," so carelessly talked of in the first scene of the play, is the very activity Edmund—the product of that "good sport"—is involved in at the moment of his father's blinding. The moment also anticipates Lear's later exclamation, "let copulation thrive," a statement that strikes the spectator of the film with all its horrible irony, as the image of Goneril and Edmund remains vivid in our minds. Moreover, there is great significance in what Kozintsev chooses not to show as he builds the two lines of action. Just as we never actually see the eye-plucking, we also never see the sexual act; Kozintsev leaves the enactment to the imagination of the spectator. Even more significant than the spur to imaginative activity to complete the images as separate events is the incentive to bring them together. By leaving the two events visually incomplete (though conceptually whole), the filmmaker makes them reciprocal agents of definition. The idea of the sexual liaison of two figures serves as the visual substance that "stands in" for the plucking out of a man's eyes. The notion of the copulating activity of a bastard son becomes the image of revenge against a father who never truly saw the consequences of his own history. The close-up on the smug and self-satisfied Edmund at the moment of Gloucester's blinding connotes the father's role in his own demise bringing to the event past deeds and acts of oppression; the image is simultaneously a visualization of Edmund's redefinition of "legitimacy," the way that he "stands up for bastards." Gloucester's cries find their destination in the place where political and sexual power join in an act of revenge for the past and a lust for power in the future.

An imaginative process of reciprocal definition occurs in the experience of film in which levels of action, if presented as a temporal equation, unfold. In Brook's film, Lear's unruly behavior displaces the revolt of Goneril, and vice versa. In Kozintsev's *Lear*, Edmund's sexual act resonates with sexual acts of the past and with the cruel act of the present; the eye-plucking in turn is the terrifying "voice" of a dark consummation. Similarly, the images of the human-made storm within graft themselves onto the spectator's experience of the "natural" storm without. In cinema, the ghost of King Hamlet *is* the panicked rearing of wild horses as much as it is the towering giant.

Conversely, the horses are the first visible form of that ghostly presence. Simultaneous performance defines a specific region of the temporal field in film and activates the imaginative process of grafting and substituting multiple action.

III

In the above examples, simultaneous action comes about through the articulation of a moment or sequence in two or more different locations. The filmmaker creates the temporal equation through editing, and the spectator reads the multiple images of the moment as concurrent action. But simultaneous action is not only a product of editing; two very distinct synchronous activities can form a dynamic through the locations of the visual and aural. Indeed the juxtaposition of things seen and things heard can form, in and of itself, a temporal equation.

The first two soliloquies in Kozintsev's *Hamlet* serve as examples. In both cases, the filmmaker creates a specific juxtaposition of the aural and visual in an effort to articulate the simultaneous events of the hero's thoughts (in Kozintsev, thinking *is* an event) and the activities in the environment out of which those thoughts emerge. Kozintsev directs Hamlet's expressions of private thought as spontaneous events of the mind, inspired by the context he finds himself in.

> Monologues . . . are not speeches but currents of thought. The inner world of the man becomes audible. From the chaos of sensation, ideas are formed. They are still in movement; no sediment has formed.[14]

The word play of "*still* in *movement*" (which is relevant to the movement–stasis juxtaposition in the film as a whole) is significant to the way Kozintsev sees the soliloquies in context. More pertinent to this discussion, however, is the sense of simultaneity Kozintsev creates in the soliloquies of *Hamlet* and his concern to establish that Hamlet's words are not premeditated philosophical tracts (as many modern productions seem to indicate) but spring from the spontaneous inner responses of the character.

We listen to Hamlet's first soliloquy (beginning, in the film, with "How weary, stale, flat, and unprofitable . . . ") in voice-over as he moves amidst the courtiers and servants preparing for the marriage festivities. In the film, Hamlet wanders through the celebration of an "o'er hasty marriage." His internal thought exists in a dynamic with the attempts of the current political regime to cover its own corruption, to create an appearance of harmony and good cheer. Whether the spectator recognizes the specifics of the corruption at this point in the film is not important: we still know that the King has just recently died; the Queen has married her brother-in-law; Hamlet, the son, strangely, has not inherited the throne and is the only one in mourning. We may not know the specifics of the problem, but we certainly recognize that a boisterous celebration is ill-timed.

The simultaneity of the aural and visual works to clarify and announce in a political context the nature of the events before us. With the servants milling about in preparation we see what, to Hamlet, is the "unweeded garden" he speaks of, as well as that which is "rank and gross in nature"— the human beings of Elsinore. By presenting the soliloquy as one element in a mode of simultaneous action, Kozintsev offers a specific performance context for us to understand the source of the hero's sexual nausea and dismay. On the other hand, by keeping the soliloquy in the covert space of the voice-over, the filmmaker articulates Hamlet's independence, his refusal to accept Claudius's reign passively, and his threatening role in Denmark:

> The interior monologue [is] particularly interesting if it is successful in giving the impression of an explosive force of thought which betokens danger for the government of Claudius. Spies have instructions to shadow this dangerous man, and not to let him out of their sight. And Hamlet unhurriedly and calmly strolls about the room. The camera goes closer; we hear the words of his thoughts, but the sleuth who clings to the door hears nothing. . . .[15]

We recognize the "danger" of his thinking because we understand his soliloquy in relationship to all that surrounds him. Through the simultaneous action of the visual and aural fields, Kozintsev articulates two forces at war in the drama: the regime of Claudius, with its ostentatious seeming and obsequious, politic worms, and Hamlet's refusal to capitulate to those in power. Fascism cannot accommodate the scrutiny of a single man—tyranny and freedom of thought collide in a moment of performance. Kozintsev politicizes the soliloquy and, in the process, offers a sense of hope: Hamlet is potentially the force that will expose the hypocrisy of the human beings in Elsinore.

Kozintsev presents Hamlet's soliloquy in the second act ("O, what a rogue and peasant slave am I!") in a similar manner. The scene with the players (2.2) takes place outdoors in a courtyard. When Hamlet goes to greet them there is a great flourish of activity. Music, tumbling, shouting, and dancing provide an atmosphere of mirth in a film that, until this point, has been largely melancholic. Hamlet interacts with the players in a playful and spirited mood. When the leading player finally recites Aeneas' tale to Dido, Hamlet sits on the edge of a prop wagon, amidst all the theatrical accoutrements of the troupe. Hamlet among the props constitutes a relationship of significance that expresses the hero's dilemma: he refuses to be a pipe to be played upon, and yet, in context, he feels the pressing reality of being just another theatrical prop; hence his frustration as a "rogue and peasant slave." On the words of the First Player, "The instant burst of clamor that she made," Hamlet begins to beat a small drum in a regular, hypnotic rhythm. The sounds of the drum provide a haunting prelude to the anxieties of the hero. Towards the end of the Hecuba speech the camera focuses on Hamlet as his soliloquy begins in voice-over. Hamlet's thought and the theatrical event of the players unfold in a relationship of simultaneous action. His soliloquy overlaps with the end of the player's monologue in the film (Hamlet's words take over on the soundtrack even though the player continues). Hamlet chastises himself:

> Who calls me villain? Breaks my pate across?
> Plucks off my beard, and blows it in my face?
> Tweaks me by the nose? Gives me the lie i' th' throat,
> As deep as to the lungs? Who does me this?
> Ha, 'swounds, I should take it; for it cannot be
> But I am pigeon-liver'd, and lack gall
> To make oppression bitter, or ere this
> I should have fatted all the region kites
> With this slave's offal. Bloody, bawdy villain!
>
> (2.2.572–80)

To facilitate the soliloquy on stage, Shakespeare has the players exit; but in the film the specific dramatic context in which we find the hero catalyzes, in an immediate sense, the hero's thoughts. At moments when the First Player speaks, the camera scrutinizes Hamlet; at moments when Hamlet "speaks" (in voice-over), the camera shows the players and their theatrical equipment. We are in two places simultaneously. Like the grains of Zeno's paradox, film divides the single performance beat, this time stimulated by the juxtaposition of the aural and visual. While the sound (the language) leads us to recall the speaker who is not in sight (thus modifying expectations for our next glimpse of him), and while what we see (the silent witnesses of props and players) exposes what we cannot hear (also complicating the "shading in" process of what is to come), one recognizes the viewing process as an imaginative dynamic of temporality. Hamlet's frustration with himself, his recognition of the significance of the theatrical event before him as a comment on his own circumstance, his struggle not to capitulate, and his feeling that his behavior shows him to be just another theatrical prop—all find expression in the film through his spontaneous reaction to the player's speech. On the other hand, the speech of the player that we hear on the soundtrack, while the visuals focus on Hamlet, serves as an important filter through which we perceive the hero and understand his thoughts. At one moment during the soliloquy, we see Hamlet touch his forehead as if the thoughts that are going through his mind are about to explode. Because we hear the speech in voice-over, the words themselves seem imprisoned in

Hamlet's mind and lack even the release of direct speech. The voice-over technique at this moment yields the sense of how thoughts themselves are an overwhelming force, trapped in the consciousness of the hero. Hamlet resolves to convert the whirling energy of his thoughts into practical action and decides upon the play as a way to test the truth of the Ghost and Claudius's culpability. Shostakovitch's music swells, and, at the end of the speech, Hamlet's only outward expression is a wordless scream.

IV

Two soliloquies of Kozintsev's *Hamlet* unfold in the simultaneous action of the aural and visual fields. In the one instance, Hamlet's thoughts take on a specifically political meaning in context, while in the other the accoutrements of the players in combination with the speech on Aeneas invade the hero's consciousness and become the focal point of his own self-criticism. In *King Lear*, on the other hand, Kozintsev works with simultaneous action in a different manner. Simultaneity is at the base of the conceptual focus of a film that maintains a fundamental tension between despair and hope, between destruction and the incomprehensible human will to rebuild. Extremes of despondency and optimism inform the film from the moment of Gloucester's blinding, and, through his imagery and moments of simultaneous action, the director not only builds that tension but finds a way to demonstrate what he sees as a final triumph of hope in Shakespeare's tragedy.

The element of hope at the foundation of Kozintsev's work parallels a pattern he discerns in the play itself. Out of a terrifying portrait of human suffering, a spark of hope can still emerge. One thinks of the servant who defends Gloucester, of the reconciliation of Lear and Cordelia, of Kent's loyalty, of Edgar's compassion. In Kozintsev's words, "a simple digit falls off from the majority. A digit of goodness, truth, compassion—against the mass of evil, falsehood and cruelty." Moreover, "an old man, seriously ill, came to his senses (he was senseless for a long time)."[16] Even in a production as pessimistic and nihilistic as the one directed by Peter Brook, Kozintsev still felt an optimistic light in Shakespeare's text shining through. Commenting on Brook's work, Kozintsev writes that "in a production which emphasized hopelessness, hope triumphed."[17]

In Kozintsev's film, after evil reaches its height (through the blinding of Gloucester and the union of Goneril and Edmund), it seems that Armageddon has arrived. But simultaneously with images of destruction, a new spirit emerges. The first shot we see after the terrifying scenes in Gloucester's castle is of Edgar taking clothes off a scarecrow; we listen to his soliloquy in voice-over, and the words set the dominant tone for the rest of the film:

> Yet better thus, and known to be contemn'd,
> Than still contemn'd and flatter'd. To be worst,
> The lowest and most dejected thing of fortune,
> Stands still in esperance, lives not in fear.
> The lamentable change is from the best;
> The worst returns to laughter.
>
> (4.1.1–6)

In these few words, Edgar seems to sum up the point of Kozintsev's film. The scenes of destruction and horror that dominate the last third of the film are not presented as statements on the meaninglessness of life. On the contrary, Kozintsev shows how "esperance" continually manifests itself amid the disaster. After Edgar's soliloquy on hope, we watch, in close-up, an old man tenderly nurse the eyes of Gloucester. Poor Tom, the beggar, then becomes the blind man's guide in the grey and desolate wasteland. Peasants and beggars continue to roam in the background, and Edgar and Gloucester become one with them. Suddenly we hear again the cry of the old, primitive horn with the accompanying visual "refrain" of wandering. Images of lonely, exhausted, desolate beggars wandering a sterile landscape emerge simultaneously with the sound of the horn, a sound that calls for community, for a fight against a wave of destruction seemingly impossible to halt. It is the sound of hope for humankind.

The next time we see Lear, he is crawling through the grass picking flowers. The camera is low to the ground, emphasizing the depths the King has fallen to. He gnaws at roots in search of some sustenance. The King is now one with the beggars and peasants who surround him; they, too, search the wasteland for some food and water. For a third time, we hear the cry of the horn, and Lear, now with Gloucester and Edgar, accepts the call, for on the King's line "a man may see how this world goes with no eyes," the three men begin to wander with the horde of beggars. Kozintsev shoots the meeting of the two old men, now the lowest things of fortune, in the context of the wandering masses, masses they join in a response to this mysteriously powerful call.

From the moment the French land on English soil (4.4), anarchy permeates the film. The dominant images are those of armor, weapons, and soldiers wandering in their battalions. Fire is everywhere. We see villages and fortresses destroyed. The depth of human suffering is made evident by Kozintsev's unrelenting visual portrayal of a world in war. Unlike Welles's depiction of the battle of Shrewsbury, however, the focus is not on the actual conflict of men in arms, but only on the results of such conflict. While Welles portrays the process of war that leads to destruction, Kozintsev shows only the consequences:

In *Lear* it is not the panoramic events which are important—the movement of battalions, advances, attacks—but faces and eyes: if you look deeper into them you can see (in their eyes alone!) the fields of battle, the defense of fortresses and the piles of bodies.[18]

Again, as with his treatment of the blinding of Gloucester, Kozintsev uses his camera, not for realistic detail, but for a visualization of the effect that the heinous crimes of men have on humanity itself.

Simultaneously with this picture of human suffering emerges (consistent with the general pattern of the film) a glimmer of hope in the reconciliation of Lear and Cordelia. Immediately before Cordelia wakes her father, the Doctor calls the Fool to play on his flute. Kozintsev keeps the Fool in the film as an important element in the pattern of hope emerging out of apparent hopelessness. Lear lies asleep on a bed of hay in an old wooden cart. Cordelia is on her knees beside him. Concurrent with this are the background activities of war, articulating, in a powerful moment of dramatic action, the astonishing range of human action in this tragedy. Cordelia begins to speak in full-frame close-up:

CORDELIA: How does my royal lord? How fares your Majesty?
LEAR: You do me wrong to take me out o' th' grave.
         Thou art a soul in bliss; but I am bound
         Upon a wheel of fire, that mine own tears
         Do scald like molten lead.

(4.6.45–49)

While Lear speaks, the full-frame shot of his face reinforces everything this film has led to. His face is worn, every line telling his story. His white hair blends with the hay on which his head lies. His eyes are exhausted. He appears the color of the wasteland itself. Suddenly a flourish interrupts the scene, and soldiers run to battle. The clanging of armor and the gallop of horses replace the sad sounds of the Fool's flute. Lear and Cordelia cannot remain still. The reality of this world is still wandering, and so they join the refugees in movement. Lear says his "forget and forgive" speech on the move, simultaneously with the images of war's appalling destruction; though his words are to Cordelia, the visuals of the suffering masses in the background make these words a statement to all humanity.

To articulate the concurrent reconciliations of parent and child and to complete the parallel between plot and subplot, Kozintsev cuts to a shot of Edgar and Gloucester roaming on a desolate plain. In a long shot, they appear alone and small. The camera cranes to an aerial perspective as we watch Gloucester stumble and fall. A close-up on Edgar's face follows. His father "scans" the face of his son with his hands. With a gasp, Gloucester recognizes Edgar. But with his next breath he dies as his heart, " 'Twixt two extremes of passion, joy and grief, / Burst smilingly." At the lowest point of existence, in the grey, lifeless space of tragedy, father recognizes son and a glimmer of joy momentarily gives light to the world.

Against the background of Shostakovitch's thundering choral score, a sequence of images represents a universe on the brink of destruction. Indeed, in the next few moments of the film, we seem to arrive at nothing short of Armageddon. Edmund sets fire to the straw huts of a village. In montage, we first see Regan and Goneril wandering amid the destruction in search of Edmund. The camera then scans the land to reveal, in quick snippets, men fighting and falling to their death. Suddenly a herd of bulls breaks out of its burning stable in a wild stampede. Simultaneously, Edmund and Goneril come together, embrace and kiss amid the ruins; this scene of destruction is the appropriate setting for their lust. We see mobs of refugees fleeing in search of safety. Mothers hold their children and run with the sick and wounded. Everything is on fire. Human beings destroy the world.

Edmund succeeds in capturing the French army, and soldiers are brought before him as prisoners, filing past him, relinquishing their weapons. Suddenly, out of the phalanx, Lear and Cordelia appear. Under heavy guard, soldiers escort them to prison; but although we see the guards around Lear and Cordelia in the visual field (as well as the activities of the war in the background), Kozintsev cuts out all sounds of battle to isolate in the aural field a soft choral chant behind the King's words:

> No, no, no, no! Come, let's away to prison.
> We two alone will sing like birds i' th' cage.
> When thou dost ask me blessing, I'll kneel down,
> And ask of thee forgiveness. So we'll live,
> And pray, and sing, and tell old tales, and laugh
> At gilded butterflies, . . .
>
> (5.3.8–13)

Through the simultaneous action of a visual field of war and an aural field of a king's gentle words of humility, we again recognize the miracle of esperance in the midst of annihilation. The spectator enters, briefly, into a parenthesis of happiness within the context of destruction. By cutting out all external sounds, Kozintsev creates for Lear and Cordelia (and for the audience) a protected moment of peace and serenity. We do not forget that Edmund still wields his weapons of destruction (in fact we see havoc and ruin all around); we simply find momentary release as we share the joy between father and child.

Defeated by his avenging brother, Edmund grovels in the mud and tells of his order to "hang Cordelia in the prison." Traditionally, in a stage production, Lear enters at this moment with the dead Cordelia in his arms (5.3.259). But in the film, the King does not enter the scene physically;

rather, the camera seems to seek out Lear at the peak of his misery. It lingers on the dying Edmund, and, in one of the most moving moments of any Shakespeare film, we hear on the soundtrack the howls of Lear echoing through the desolate land. The cry of the horn yields to the cry of the human voice. In the visual field we see Edmund; in the aural field we hear Lear's cry. Incredibly, that sound, even given the bloodthirsty aberration of humanity that is Edmund in this film, is a sound of hope; it is astonishing that this man can still cry and is not numb with pain.

> The cry of grief, bursting through the dumbness of the ages, through the deafness of time, must be heard. We made the film with the very purpose that it should be heard.[19]

The camera finally pans to the top of the battlements and, from the perspective of those below, we see the tiny figure of Lear above, silhouetted against the sky. To the King's left, in an arch on top of a mountain cliff, hangs Cordelia. The film cuts to a shot from behind Lear looking down at the masses of soldiers below. The King cries out, "O, you are men of stones!" Reverberating on the soundtrack, Lear's voice echoes with his agony as he wonders that those around him do not crack "heaven's vault" with their grief.

A soldier cuts Cordelia's body from the rope and lays her on the ground. In a close-up, Lear kneels by the body, begging for her life. Kozintsev refuses to dominate the last moment with visuals that would determine whether the King dies in joy or despair. He pulls his camera back to reveal a panoramic shot of Lear by the body, but father and daughter are blocked by Kent, Albany, and Edgar surrounding them. Visual and aural fields are distinct once again and we listen to Shakespeare's words but fail now to see the object of those words. "Do you see this? Look on her, look, her lips, / Look there, look there!" Does Lear believe her to be alive? Or is this final moment simply a last glance at those lips that once dared to utter the truth? The ambiguity remains. Kozintsev will not offer an answer.

In the last moment of the film, we again witness the pattern of hope and optimism coming through apparent despair. Soldiers carry the bodies of Cordelia and Lear on a litter. The camera tracks backwards to show the Fool on the road and in the way. One of the soldiers kicks the weeping Fool aside. At first glance, one feels as if this is a final note of pessimism in Kozintsev's film. In fact, however, this is not the last of the Fool. Recovering from this brutality, he sits up and begins to play a tune on his flute. A human being bounces back from the pains of this world. The Fool plays music. He creates.

> [The Fool's] voice, the voice of the home-made pipe, begins and ends this story; the sad, human voice of art.[20]

CHAPTER 7

Naming Time: Orson Welles's *Othello*[1]

Orson Welles's *Othello* is a Shakespeare film distinguished by a complex temporal strategy that offers, with remarkable power, penetrating insights into the dramaturgical design of this tragedy. Welles's considerable knowledge of Shakespeare and his talents as a filmmaker combine to produce a film interesting not only for its capacity to present the play as a function of the cinematic medium, but, also as a critical essay on *Othello* in its own right. Indeed, Welles' conceptual focus on the nature of time in the play, and the specific way he explores temporality through cinematic technique, demonstrates that film can serve as a critical tool for reinterpreting Shakespeare's work. Benjamin's "new field of perception" becomes (strikingly in this film) a new field for exploring the specific relationship between the text and the world that the film opens as it draws from that text. Because of its uniqueness as a study of time in *Othello* (and the way the play becomes specifically a function of cinematic time), Welles's film is the sole focus of this chapter. I hope to make explicit en route what has been an implicit part of this study from the beginning: Shakespeare on film holds an important place in a larger critical tradition.

I

Welles spent the years from 1948 to 1952 producing an *Othello* that, despite its technical flaws, is a film of extraordinary visual beauty. After completing the film he remarked,

> In *Othello* I felt that I had to choose between filming the play or continuing my own line of experimentation in adapting Shakespeare quite freely to the cinema form. . . . *Othello* the movie, I hope, is first and foremost a motion picture.[2]

Welles's presentation of Venice, with its stately buildings, its calm and channeled waterways, and its solid appearance, reflects well the sense of order achieved—temporarily—in the first act of Shakespeare's drama. Moreover, the visuals of the Venetian world serve as a harmonious complement to the nobility and stature of the hero before the "green-eyed monster" of jealousy overwhelms him. In juxtaposition with the ordered world of Venice is Cyprus, with its jungles of arcades and pillars, its seamy underground, its narrow and winding streets, its stairways, its high and frightening cliffs and battlements, and its roaring ocean shore. If Venice is the setting that corresponds to Othello, Cyprus is the complement to Iago. In this world the villain reigns supreme, and he uses the twisted and confusing dimensions of the Cyprus environment to create an unrelenting hell for his victims.

As the film progresses the spectator realizes how the *mise-en-scène* of the film is a product of the director's sensitivity to the conceptual operation of time in the text. The guiding force of temporality is evident from the prologue. The film opens with the funeral processions of both Othello and Desdemona silhouetted against the sky. Othello is carried on a bier followed by a long line of monks in black while Desdemona is carried and followed by a line of monks in white. Although the processions are clearly separated (physically and through color), the rhythm of the two lines is synchronized. Bells chime in a regular beat throughout the prologue, suggesting a firm and cohesive sense of time. The pace of the processions (visually) and the repetition of the accompanying chimes (aurally) yield the sense of unity through compatible rhythm. As in many of Welles's films, *Othello* opens with visual images taken from the very end of the film, not because the director wants to remove all suspense, but because he wants to establish a unified sequence for the whole work. In other words, the way the plot unravels takes precedence over the surprises of narrative. Moreover, by using the dominant rhythms of the processions to establish a sense of coherence and order early in the film, Welles can then illustrate how Iago shatters this tight temporal structure to bring chaos into the world of the play and its hero. Indeed, when one sees the processional image again at the end of *Othello*, one recognizes how the rhythm of time has been broken and restored through the course of the tragedy.

In the opening sequence, Welles intersplices the funeral processions of Othello and Desdemona with shots of Iago dragged by chains through a crowd of screaming Cypriots. Guards throw him into an iron cage and haul him to the top of the castle walls. We witness the world momentarily from Iago's perspective; the cage spins as it hangs, the crowd screams, and, as long as we are with Iago, we lose the stately rhythm of the processions. In this prologue, Welles develops his temporal theme by realizing the rhythms of Othello and Desdemona on the one hand and Iago on the other. William Johnson sees the entire film in a structure of contrasting rhythms, and his sensitivity to this aspect of Welles's work is rare. In the following passage he seems to scan the rhythms of the film.

Orson Welles's *Othello*. Michael MacLiammoir (center) as Iago. (*Courtesy of the Academy of Motion Picture Arts and Sciences*)

> Welles sets the whole tragedy in perspective with an opening sequence that
> interweaves the funeral corteges of Othello and Desdemona and the dragging
> of Iago to his punishment. . . . [But] the staccato rhythm associated with Iago
> gradually imposes itself on Othello's stately rhythm, and the increasing com-
> plexity of the film's movements suggests the increasing turmoil of doubt in
> Othello's mind.[3]

Welles uses the rhythms of time to guide the spectator through *Othello*.
The film conveys an objective sense of time in the aural field through the
regular pattern of beating drums or chiming bells. With this use of the
soundtrack, the events of the drama (and its characters) function within a
contained and steady passage of time. At certain key moments, for example,
the spectator hears the footfalls of characters in a regular and constant
rhythm, yielding the sense of the individual's participation in time's inevitable course. Like the ticking of a bomb about to explode, however, these
regular patterns erupt in corresponding sounds and images of chaos as
realized by the tempest, the crashing waves of the Cyprus shore, the sudden
explosion of cannon fire, the wild break of seagulls in the sky, and the
uncontrolled and chaotic revels after the defeat of the Turks. In an objective

sense, one recognizes an ambiguous sense of time—ordered and chaotic, constant and fragmented.

I use the term "objective" only to differentiate the presence of time as a force in the film from individual relationships to that force. For it is in Welles's development of the subjective experiences of time, his exploration of the temporality of character, that he works out his conceptual focus most effectively. On the one hand, we understand time through Othello's experience: what is clear and chronologically sound in the first part of the film eventually becomes distorted and irregular as the drama progresses. As Othello's pain and jealousy increase, we lose a sense of coherence in the film. Through the unique resources of the cinema—aural and visual—Welles realizes the hero's experience as he is overcome by jealousy. Early in the film the director shoots Othello in clear light, but as the film progresses we see him increasingly in shadow. Through the use of montage, our sense of space and time disintegrates. Harsh, dissonant sounds eventually replace the regular and even sounds of the first part of the film. As jealousy and madness overwhelm the hero, we watch him traverse the spectrum from order to chaos, from light to shadow, and, as a result, we understand how Iago has set out to destroy his victim. He causes Othello to see only the dark, chaotic side of time—something that the hero fears and that is fundamentally against his character. "And when I love thee not," he says of Desdemona, "chaos is come again." Chaos represents a movement backwards for the hero, a state without love, a destruction of his sense of the eternal.

By contrast, Iago perceives time as an agent to control. He emerges as the master of time in the film, and the "success" of his scheme relates to his ability to manipulate, not only the objective force of time, but also Othello's relationship to that force. Welles develops this idea early in the film. Still in Venice, Iago is working on his gull, Roderigo, when he comes forward to the camera (recall Richard's similar boldness with the camera) and says, in close-up, to the audience, "I am not what I am." Immediately following his words the scene dissolves to a close-up shot of mechanical figures striking the bells of the clock in St. Mark's Square. In perfect mechanical order these human impressions (the figures are human in shape but not in substance—i.e., they are not what they are) strike. Welles's use of the dissolve here forces the spectator to associate these figures with Iago as one who will hammer upon Othello's emotional balance as the figures hammer upon the chimes. But the specifics of the image also suggest that he will achieve his ends through a controlled use of time. As the "Jack of the clock"—to borrow the trope of Richard II—marks time, so, too, will Iago orchestrate his destruction of Othello. To make this metaphor clear, Welles

Orson Welles's *Othello*. Othello and Desdemona in the world of Cyprus—the concrete net to "enmesh them all." (*Courtesy of the Academy of Motion Picture Arts and Sciences*)

follows the shot of the mechanical figures with a dissolve to the bedchamber of Othello and Desdemona. Othello parts the curtains surrounding the bed, and, in a high-angle shot, we see Desdemona lying on the bed with her long blonde hair spread out underneath her. Othello then speaks his lines from Act 1, scene 3:

> Come Desdemona. I have but an hour
> Of love . . .
> To spend with thee. We must obey the time.
> (1.3.301–303)

Eventually one recognizes Othello's words "we must obey the time" as highly ironic because an obedience to time, in this film, translates into an obedience to the one who controls time—Iago. The hero bends down to kiss his bride, followed by a dissolve to black. In this sequence of images Welles encapsulates the entire drama: Othello and Desdemona "obey" time as it is orchestrated from without but, ultimately, this obedience leads to an overwhelming blackness, consuming beauty and love.

II

Welles's film inspires an important question: How does time function as a dramaturgical device in *Othello*? Because the film led me to ask this question, and because Welles's treatment ultimately inspired a fresh reading of the play and exposed an aspect of the text I had not sufficiently noted, I pause here to return to the text to examine in some detail Shakespeare's own exploration of time. I delay a detailed treatment of how time works as a focal point in the film to clarify the function of time in the text.

Iago's words to Roderigo in the second act reveal his skill in manipulating time:

> Thou know'st we work by wit, and not by witchcraft;
> And wit depends on dilatory time.
>
> (2.3.366–67)

A review of Iago's speeches demonstrates that they are filled with the word "time" and that his language often uses temporal imagery. For example, in the very first scene he is angry because time has failed to bring about his expected promotion. He begins by speaking of his frustrations in Othello's service and of his jealousy of Cassio:

> This counter-caster,
> He, in good time, must his lieutenant be,
> And I—God bless the mark!—his Moorship's ancient.
>
> (1.1.32–34)

Iago believes that he has proved himself in time but still has lost the promotion. Preferment seems to have nothing to do with loyalty in time:

> Preferment goes by letter and affection,
> And not by old gradation, where each second
> Stood heir to the first.
>
> (1.1.37–39)

Like Richard II, Iago believes that he has wasted time; but unlike the King, he will not allow time to waste him. He will not be like

> Many a duteous and knee-crooking knave
> That, doting on his own obsequious bondage,
> Wears out his time, much like his master's ass . . .
>
> (1.1.46–48)

Iago then calls for a bell to wake Brabantio (an image akin to the mechanical figures who mark time in Welles's film). He begins here to use time actively to achieve his ends.

Iago's perception of time is close to Machiavelli's notion of Fortune in *The*

Orson Welles's *Othello*. Orson Welles and Suzanne Cloutier as Othello and Desdemona. (*Courtesy of the Academy of Motion Picture Arts and Sciences*)

Prince.[4] For Machiavelli, Fortune is like a river that fluctuates between extremes of chaos and peace. But, he says, one can take precautions by building "floodgates and embankments" in quiet times so that the violent times can be controlled. One of Machiavelli's central points in *The Prince* is his call for an easing of the control of Fortune in human affairs. If one exercises the highest Machiavellian virtue of prudence, one can learn to take control of much of one's destiny.

Iago emerges as a figure who works with time in the manner that Machiavelli prescribes. In the first scene of the play Iago tells us that he had waited for time to give him his promotion but was ultimately frustrated. Now, however, he will conquer time by taking control. Machiavelli's metaphor of the river is akin to Iago's use of images associated with pregnancy, gestation, and birth. For example, in a discussion with Roderigo towards the end of the first act, Iago tells his gull that "There are many events in the womb of time which will be delivered" (1.3.371). In the soliloquy that ends the scene he concludes,

> It is engendr'd. Hell and night
> Must bring this monstrous birth to the world's light.
> (1.3.404–405)

Indeed, Iago fertilizes time so that it will give birth to his desires. At a propitious moment, he will inject his poison into Othello with tales of Cassio and Desdemona: "After some time to abuse Othello's ears / That he is too familiar with his wife" (1.3.396–97). From this point in the play he will work so that, to use his words, "Time shall . . . favorably minister" his ends.

When next we hear of Iago we discover that he has indeed begun to conquer time. On the shores of Cyprus, the Second Gentleman remarks that Iago has fought his way through the tempest and has landed safely. Cassio comments on how the ancient (ensign) has had "favorable and happy speed" (2.1.68). He comments approximately ten lines later that Iago has defied the expected time of his arrival. Desdemona was "Left in the conduct of the bold Iago, / Whose footing here anticipates our thoughts / A se'nnights speed" (2.1.76–78). It is interesting to note that at this point in the play Shakespeare juxtaposes Iago's victory over time with Othello's corresponding delay.

As mentioned earlier, Iago's speeches are filled not only with many utterances of the word "time" but with images of temporality as well. Curiously, Iago's victims use this word and corresponding images with increased frequency as the villain gains control. G. Wilson Knight points out that Othello eventually enters Iago's "semantic sphere," and Jan Kott argues, "Not only shall Othello crawl at Iago's feet; he shall talk in his language."[5] But it is in their specific concern with issues of time and their use of temporal imagery that Iago's victims interest us here. For example, when Othello discovers Cassio, in Iago's words, in a "time of his infirmity," he dismisses the lieutenant from his post. Interestingly, Cassio confides in Iago that what bothers him most is not that he has disillusioned the general but that his reputation will suffer. He defines that reputation as "the immortal part of myself." In a sense, Cassio perceives that he has lost time by losing his post. Iago "comforts" Cassio by telling him that nothing is final—that, in time, he can regain his lieutenancy. After all, "you or any man may be drunk at a time, man" (2.3.307). What is most significant here is that Shakespeare expresses Iago's control by showing him to be the master of time. It is when his victims are concerned, somehow, with issues of time that they are most vulnerable to him.

Shakespeare also represents vulnerability through temporal concerns in Desdemona's plea for Cassio. The former lieutenant asks Desdemona, in essence, to redeem time for him. She promises that she will speak with her husband and "tame and talk him out of patience" (3.3.23). When she makes her request to Othello, she insists on knowing the time when he will restore Cassio.

DESDEMONA: I prithee, call him back.
OTHELLO: Went he hence now?
DESDEMONA: Yes, faith, so humbled
           That he hath left part of his grief with me
           To suffer with him. Good love, call him back.
OTHELLO: Not now, sweet Desdemon; some other time.
DESDEMONA: But shall't be shortly?
OTHELLO: The sooner, sweet, for you.
DESDEMONA: Shall't be tonight at supper?
OTHELLO: No, not tonight.
DESDEMONA: Tomorrow dinner, then?
OTHELLO: I shall not dine at home;
           I meet the captains at the citadel.
DESDEMONA: Why, then, tomorrow night, or Tuesday morn,
           On Tuesday noon, or night, on Wednesday morn.
           I prithee, name the time, but let it not
           Exceed three days.

                                      (3.3.51–63)

Ironically, Desdemona's insistence on naming the time is a tacit victory for the one who conquers time—Iago. She becomes an unwitting accomplice in the villain's scheme. Shakespeare's clear emphasis on naming time is a comment on the vulnerability of Iago's victims. Moreover, the great "temptation scene" follows this sequence; now that his victims are working on his terms and using his language, Iago finds the appropriate moment to strike.

Iago's mastery over time manifests itself in a more fundamental and obvious way in Shakespeare's *Othello*. Simply put: he has superb timing and knows exactly when to strike. Like Machiavelli's brightest examples of political success (Moses, Cyrus, Romulus, Theseus), Iago receives nothing from Fortune but the occasion; and when that occasion arises, he makes optimum use of it. Emilia happens to pass by when Desdemona drops the handkerchief. Bianca conveniently arrives when Iago is with Cassio as Othello secretly watches and receives the "ocular proof." Roderigo happens to be in love with Desdemona and is stupid enough to abide by Iago's demands to achieve his desired ends. In each instance, Iago is able to exploit to the fullest the opportunities that his good fortune provides. Whether one is directing the play for the stage or for film, one must create situations that reveal Iago's understanding of how to take the greatest advantage of the circumstances of the moment. One can imagine the myriad ways of staging Iago's famous "Ha, I like not that," which subtly sets off the spark of destruction in Othello's consciousness. In whatever way it is performed, however, a director must present Iago's brilliant sense of timing at that moment. As we witness Iago's clever machinations, we are watching one of the keenest manipulators of time in the history of drama. He knows when to push and when to hold back. He knows when to further his wretched lie and when to keep silent. He knows how to build his "evidence" and how to bring his victim to the point where he perceives what Iago wants him to perceive. And he does all this by recognizing when the opportune moments occur. Iago knows how to work with time so that time works for him.

Iago's perception of time is thus one that assumes constant change and mutability. It cannot be trusted to bring about one's desires because, like Machiavelli's river, it is fickle and erratic. For Iago, time is the "fashionable host" that Ulysses speaks of in *Troilus and Cressida*. Indeed, to him as to Ulysses, "beauty, wit, / High birth, vigor of bone, desert in service, / Love, friendship, charity, are subjects all / To envious and calumniating time." But Iago uses the fickleness of time to his own advantage; his skill lies in his ability to exploit "calumniating time" through dissembling, plotting, and Machiavellian prudence.

In contrast to Iago's perception of time as something that fluctuates and that ultimately must be conquered is Othello's perception of time. For the hero, time has meaning and significance in its range and continuity. Unlike Iago, Othello perceives the world in terms of the eternal. His speeches seem to resound with the words "never" and "forever." He asserts that he received his "life and being / From men of royal siege." He invests a great deal in a handkerchief given him by his mother because it represents history and links him with his past and his heritage. Othello's world is based on loyalty, history, and a sense of being rooted in time. When he vows anything, from his marriage to his vengeance, his words have an everlasting implication. To Iago he swears, "I am bound to thee forever." In the words of G. Wilson Knight, "from the first to the last he loves his own romantic history."[6] His association with the grand spectrum of time in human history also leaves one with the impression of an integrated man whose nature, to borrow A. C. Bradley's phrase, "is all of one piece."[7] Iago, who knows his target, refers to Othello as one with a "*constant*, loving and noble nature."

At a key moment in the play, Othello becomes so overwhelmed with Desdemona that he expresses a desire to immortalize his love for her. As Derek Traversi points out, when the hero arrives in Cyprus and is reunited with Desdemona, "his one desire is to hold this moment to make it eternal."[8]

> If it were now to die,
> 'Twere now to be most happy; for I fear,
> My soul hath her content so absolute
> That not another comfort like to this
> Succeeds in unknown fate.
>
> (2.1.186–190)

But Othello's belief in the possibility of absolute happiness through love also reveals his greatest point of vulnerability. Implicit in his words is a restless awareness that, until death, time has the power to destroy present emotion. Traversi clarifies the paradox:

> This precarious moment of happiness will never find its fellow, for the temporal process is . . . one of dissolution and decay. Only death can come between this temporary communion and its eclipse; but death, of course, implies the annihilation of the personality and the end of love.[9]

Unlike the realistic and sensible Rosalind, Othello reaches for the "ever" and not the "now" of human love. As is so often the case in Shakespeare's dramas, one's greatest desire is also the point of one's greatest vulnerability. In a way similar to Maria's exploitation of Malvolio, Iago pinpoints the spot where his revenge will "find notable cause to work" (*Twelfth Night*, 2.3.152). Othello's statement on the manner by which he could attain eternal happiness through love exposes, by implication, his fear of the power of time's mutability. This fear becomes Iago's primary target. The villain exploits the hero's hidden anxiety and shatters his greatest hope.

Iago destroys Othello by altering his perception of the nature of time. At a number of moments in the play, Desdemona is associated with images of the divine. For Othello, she becomes "a symbol of man's ideal, the supreme value of love. . . ."[10] When Iago shows Desdemona false, Othello's sense of the eternal decays in turn. As Othello succumbs to Iago, he speaks in terms of shattered time, of a broken sense of what holds his world together. I return, for a moment, to that key passage early in the "temptation" scene:

> Excellent wretch! Perdition catch my soul
> But I do love thee! And when I love thee not,
> Chaos is come again.
>
> (3.3.92–94)

Othello's reference to "chaos" obviously foreshadows the disintegration of his being. Chaos is time with neither order nor coherence; and by destroying Othello's sense of the constancy of experience, Iago brings chaos into his life. The concept of marriage, by its very definition, is something based on the notion of eternity. It is a gift of heaven, sanctioned by the eternal. As Othello says: "If she be false, O, then heaven mocks itself." When his marriage is destroyed, everything in the hero's existence seems lost in the timelessness of chaos.

The emotion of jealousy suggests a sense of time shattered. As that "green-eyed monster" "mocks the meat it feeds on," one becomes enraged because one's sense of continuity is broken. Instances of the link between jealousy and chaos are not unique to *Othello*. A similar relationship is suggested in *A Midsummer Night's Dream* when, in Act 2, Titania attributes the disorder of the natural world to the dissension between the King and Queen of fairies; moreover, that dissension—in her view—ultimately stems from "jealous Oberon" and, in a larger sense, from the "forgeries of jealousy." Similarly, Leontes's jealousy leads to a sense of time broken as he loses the rhythm of sleep and rest: "Nor night nor day no rest" (2.3.1). The only antidote to the poison of his jealous rage is sixteen years of penitence and faith. In *The Winter's Tale*, time is restored. Jealousy belongs in Iago's domain because it is part of the capricious time he is the master of. Othello's time—constant and ordered—is one in which jealousy cannot reign. As a result, he is spoken of as "one not easily jealous"; "his whole nature was indisposed to jealousy."[11] When asked whether Othello is jealous, Desdemona replies:

> Who, he? I think the sun where he was born
> Drew all such humors from him.
>
> (3.4.29–30)

It is clear in the first half of the play that Othello's nature is "made of no such baseness / As jealous creatures are." He confirms that jealousy is part of changing time and that his sense of constancy has always kept him from such dis-ease. As Desdemona associates constancy with the sun, Othello associates jealousy with the changing moon:

> Think'st thou I'd make a life of jealousy,
> To follow still the changes of the moon;
> With fresh suspicions?
>
> (3.3.183–85)

When time loses its scope and becomes changeable like the moon, the result, for Othello, is madness and chaos:

> It is the very error of the moon;
> She comes more nearer earth than she was wont,
> And makes men mad.
>
> (5.2.113–15)

As "Iago time" takes over the play, men are made mad. Shakespeare's persistent use of temporal imagery suggests that Iago's victory is one in which he uses time to break time. In other words, his ability to manipulate time in order to achieve his ends amounts to a shattering of the constancy at the base of Othello's nobility and Desdemona's virtue. Ironically, as Othello first becomes enraged with jealousy, Iago remarks:

> My Lord, I would I might entreat your honor
> To scan this thing no farther; leave it to time.
>
> (3.3.251–52)

Leaving it to time is tantamount to leaving it to Iago. Indeed, Iago's use of the term "scan" signals once again that the villain recognizes situations by their temporal organization and examines them in terms of their rhythm. As the "temptation" scene progresses, he maps out exactly what will ensue as time goes by. Finally, one recognizes Iago's victory as Othello, like Desdemona and Cassio before him, completes the pattern of Iago's victims by demonstrating a concern with time in his lamentation of lost history and reputation:

> O, now, forever
> Farewell the tranquil mind! Farewell content!
> Farewell the plumed troop, and the big wars
> That makes ambition virtue! O, farewell!
> Farewell the neighing steed, and the shrill trump,
> The spirit-stirring drum, th' ear-piercing fife,
> The royal banner, and all quality,
> Pride, pomp, and circumstance of glorious war!
> And, O you mortal engines, whose rude throats
> Th' immortal Jove's dread clamors counterfeit,
> Farewell! Othello's occupation's gone.
> (3.3.352–62)

Although Iago destroys Othello's sense of time, the hero still speaks in terms of the everlasting when he vows his revenge. Iago's success, therefore, can be understood by the way he uses Othello's unswerving will to his own advantage. He causes the hero to translate momentary circumstance (which Iago is the master of) into terms of the eternal. Othello becomes a kind of dynamo through which Iago wreaks destruction. Once he shatters the hero's sense of time through the circumstance of the moment, Othello, in turn, translates the instant into a matter for all eternity:

> No! To be once in doubt
> Is once to be resolv'd;
> (3.3.184–85)

Shakespeare has thus remained constant in his presentation of the hero, even when the foundation of Othello's life has been destroyed: "thwarted in love, his egoism will be *consistent* in revenge, decisive, irresistible. . . ."[12]

Othello is clearly a man of action. The logistics of his elopement with Desdemona as well as his success as a soldier speak of this. He recognizes the calling of a moment and does something about it. But his action is always informed by his sense of time. This idealism is the key to the kind of nobility that defines Othello: he is a man of action but always has a sense of the relevance of that action to the greater order of the world. He is the opposite of a character like Hotspur, even in his raving jealousies, because he believes that he will redeem time through his violent deed. Having shattered Othello's sense of the eternal, Iago spurs the man of action to wreak the most horrible vengeance on the one who supposedly precipitated the fall—Desdemona. But in his brutal act, he is always conscious of "the cause" as he tries to restore time. Indeed, Othello does not "chop her into messes" as he vowed he would earlier in the play. Such an act would only further the victory of chaos. Othello smothers Desdemona.

> Othello kills Desdemona in order to save the moral order, to restore love and faith. He kills Desdemona to be able to forgive her, so that the accounts be settled and the world returned to its equilibrium. . . . He desperately wants to save the meaning of life, of his life, perhaps even the meaning of the world.[13]

In the terms I have discussed in this chapter, Othello wants to restore eternity. Facing the horror of his deed, however, he speaks of all time lost—both of the constant sun and the changing moon:

> Methinks it should be now a huge eclipse
> Of sun and moon, and that th' affrighted globe
> Should yawn at alteration.
>
> (5.2.103–105)

And in the simultaneous eclipse of sun and moon we have the death of Othello and the end of Iago's terrifying reign. Time ultimately catches up to both men in one huge eclipse.

III

In Welles's film, after Othello arrives in Cyprus, the herald stands on the battlements of the castle to announce the general's order for feast and celebration (2.2), and it is within the context of the revels that Iago begins his work. Meanwhile, Welles takes us to a second bedchamber scene, where he depicts the embrace of the lovers through shadows projected on a wall. While their bodies merge in shadow we hear Othello's "If I [*sic*] were now to die, / 'Twere now to be most happy." The moment, as Welles presents it, is fundamentally ambiguous and realizes the paradox suggested above. The words express Othello's belief in the possibility of eternal happiness through his absolute love for Desdemona, but the shadows undercut the hero's sentiments and remind us of the ephemeral nature of their love and of its vulnerability to time. Through the contrasting juxtaposition of things seen and things heard, Welles represents the mutability of a love that Othello wishes to immortalize.

The film juxtaposes the peace and calm of the two lovers with the revels outside. Welles uses the resources of the cinema to present a wild and raucous celebration, whose disorder becomes an ideal environment for the manipulations of Iago. The montage of the celebrations forces one to feel lost in a maze of drunken reveling. After Iago encourages Cassio to drink, we watch Bianca unwittingly further the villain's plan by tempting the lieutenant to indulge again. We see dancing crowds, musicians, close-ups on the instruments as well as on the bottles and the flying goblets of the drunken crowd. Welles then uses montage to present people turning and dancing as we see face after face moving about in a dizzying sequence of images. The celebrants laugh uproariously. Welles intersplices all this action with the only definable, rooted element—Iago. We see him subtly manipulating the quarrel between Roderigo and Cassio. When the fight breaks out, the celebrants seem to get even wilder, and their shouts and laughter serve as the backdrop to the encounter. A menacing feeling pervades all of this, a sense of people out of control—except for Iago, who works efficiently amidst this disorder.

At the culmination of the sequence, we follow the fighting men as Cassio chases Roderigo down into the underground of the castle, where a jungle of vaults and pillars displaces a jungle of revelers. In fact, Welles uses images of depths in his film to enforce a sense of a fallen world; he also contrasts such images by placing his characters in dangerously high places when they are particularly vulnerable—Iago in his cage, Cassio and Desdemona on top of the battlements after the war with the Turks, Othello and Iago on a huge cliff overlooking the sea at the culmination of the "temptation" scene. By shooting the encounter of Cassio and Roderigo in the underworld of Cyprus, Welles creates a visual hell: a concrete net to "enmesh them all." The screams and laughter of the celebrants now echo off the walls of this underworld, increasing the eerie sense of a world gone mad. Chaos is unleashed, on Iago's inspiration. All coherence is gone.

When Othello enters, he dismisses Cassio and apparently restores order. As the gathering breaks up, Cassio laments his dismissal by focusing on his lost reputation. As mentioned earlier, Cassio's main concern is that, with his loss of reputation—"the immortal part"—he has also lost time. Immediately before Iago urges Cassio to work for his reinstatement through Desdemona, we hear the crowing of the cock—a sign that the order of time is, temporarily, restored. In Machiavelli's terms, the river of Fortune will have its wild moments and its corresponding quiet times. When things are settled one must build the floodgates and embankments necessary for the inevitable chaotic times to come. This practice is the key to mastering time and Fortune. Thus, with the signal of the crowing cock, calm is restored and Iago begins to build. Significantly, Welles uses the camera to distort Iago's shape with a low-angle shot, making the villain loom (as a physically exaggerated form) over Cassio. The "huge" Iago persuades Cassio to appeal to the "real" general of the times ("Our general's wife is now the general"). The scene concludes with the regular beating of a drum.

Welles's treatment of the "temptation" scene (3.3) reinforces Shakespeare's representation of Iago's ability to manipulate time. Whereas we have just witnessed Iago inciting chaos within the celebrations, we now witness the master of time creating chaos within Othello himself. Interestingly, Welles shoots the entire episode with a long traveling shot of the two men walking up the battlements, establishing a sense of constant movement. Welles exploits the cinema's unique capacity to perform scenes in motion, and the effect of his presentation is to create the feeling that Iago, the manipulator of time, works best on the move. Moreover, on the soundtrack we distinctly hear the regular pattern of the footfalls of the two men as they walk in perfect synchronization. When they reach the top they are, significantly, on the edge of an enormous cliff as Iago leads his victim to a state of precarious balance.

The "temptation" scene continues as the two men go inside the castle. Again, Welles takes us below into an underworld as Othello becomes more and more enraged at the prospect of Desdemona's infidelity and as his world loses its equilibrium. During the "I see you are moved" section of the dialogue (2.207 ff), Iago actually helps Othello take off his armor. What must sound like a banal and obvious image when described in words is actually a very powerful moment in the film. Iago disarms Othello, physically and spiritually, and Welles succeeds in accomplishing a strong visual statement to support this dimension of the scene. Moreover, the shots become more and more tilted as Iago tips the balance of Othello's world.

As Iago continues to work on Othello, we see, at approximately line 234 ("And yet, how nature erring from itself—"), the Moor examining himself in a mirror that seems to distort his image. As Iago shatters Othello's sense of coherence and estranges the Moor from himself, we witness the hero confronting a "shadow" of his being. But it is Iago who stands behind the mirror, making the unmistakable statement that he is the device through which Othello perceives not only the world but himself as well. The man who was once able to say "My parts, my title, and my perfect soul / Shall manifest me rightly" is now beginning to confront the possibility of "erring nature" and its distorted shadows. Othello's "parts" are no longer integrated, and the man whom Bradley saw at the beginning as "all of one piece" is now coming apart.

Shakespeare uses images of fragmentation in his play to create a sense of a world that has lost cohesion. This idea is especially apparent in Othello's repeated references to destroying Desdemona by ripping her apart. Because Iago has destroyed his sense of constancy and eternity, it follows that Othello responds with vows of revenge that are associated with disintegration. Welles echoes these images of fragmentation in his film through montage, a quick-moving camera, and the contrasting rhythms of the soundtrack.

Towards the end of the "temptation" scene, for example, Iago gains ground in his deception by telling Othello that he "lay with Cassio" and that the latter, in his sleep, spoke of his love for Desdemona. Othello becomes enraged. At this juncture, he first speaks of revenge in the short line, "I'll tear her all to pieces." Here Welles employs a powerful counterpoint to Othello's statement. During much of the "temptation" scene the waves of the sea rhythmically beat on the fortress walls. When Othello speaks his first words of revenge, a sudden boom of the crashing sea punctuates the wrath of his exclamation. To support Shakespeare's imagery of fragmentation, Welles uses the soundtrack to suggest a break of rhythm and order with the fierce explosions of the sea. Similarly, after Othello is deceived by the "ocular proof" that Iago provides in the contrived discussion with Cassio and Bianca, the hero vows, "I will chop her into messes. Cuckold me?" Here Welles uses montage as he cuts to cannon after cannon firing to announce the arrival of Lodovico's ship. As Iago destroys his sense of order, Othello responds with images of disintegration in his vows of revenge, and Welles uses cinematic technique to emphasize a corresponding transition from order to chaos in the world outside.

Following the exploding cannons that announce the arrival of the Venetian ships, Welles cuts to a shot of Iago and Othello walking rapidly along the battlements, and a sequence representing the hero's trance then follows. As the dialogue of Act 4, scene 1 is spoken, Welles uses quick cuts from one speaker to the other to produce a sense of approaching hysteria through the speed of the cuts and the rapid gait of the characters. When Iago speaks the words "Lie with her? Lie on her?" Othello runs into the concrete jungle below the castle. We see shadows of jail-like bars projected on his frantic body and hear frenzied electronic music sustained on the soundtrack. Suddenly, with no preparation, we are with Othello on top of a watchtower. Our sense of space and time is completely disoriented. The following shot is a low-angle perspective on the tower as we are now transported to a new location below. On the soundtrack we hear the heavy, rhythmic breathing of Othello, which conveys the sense of the hero's desperate attempt to hang on to the rhythm of his life with the very air that sustains him. The next shot shows seagulls wheeling in the sky, followed by a close-up of Othello's face as he lies on his back on the ground. We then assume his perspective and another low-angle shot of the tower reveals a number of people looking down and laughing at him. With the use of a reverse zoom, the faces of these frightening spectators appear tiny and extremely far away. Othello mutters something about the handkerchief during which we also hear the sound of Iago's voice crying "My Lord" as he searches for his "master." The next shot is of the tower once again, but this time the haunting faces are gone. Iago arrives on the scene, revives the hero, and the sequence ends with Othello's "Farewell" speech, transposed from Act 3, scene 3. In this sequence, Welles converts Shakespeare's stage direction *"Falls in a trance"* into cinematic values that create a sense of spatial and temporal disorientation. Through this, one identifies with Othello as his world comes apart and as Iago destroys him by sabotaging his sense of order and coherence—all culminating in Othello's farewell to his own history and occupation.

Finally, Welles works with time in one more sense that has its seed in Shakespeare's text, but he develops it in a way that adds a further dimension to his interpretation. It it the concept of time that is ultimately responsible for restoring order to the world. It is a redeeming force of time beyond human manipulation. It is the kind of time that Viola speaks of in *Twelfth Night*, time that unties the complicated knot of that comedy. It is "the old common arbitrator" that Hector refers to in *Troilus and Cressida*. In short, a form of time that is ultimately triumphant in *Othello*.

At the end of *Othello* Lodovico speaks to Montano of the incipient punishment for Iago:

> To you, Lord Governor,
> Remains the censure of this hellish villain,
> The time, the place, the torture. O, enforce it!
> (5.2.376–78)

Welles takes this reference at the end of Shakespeare's text and develops it into a motif throughout the film. It will be recalled that the film begins with the funeral processions of Othello and Desdemona and that Iago is simultaneously dragged by chains through the crowd, thrust into an iron cage, and hoisted to the top of the castle walls, where he is left to hang. From the opening of the film, therefore, we are aware of his ultimate punishment. At four key moments during the course of the film Welles shoots the empty hanging cage, reminding the audience of Iago's eventual punishment. Welles lingers with a shot of the cage during the herald's announcement before the celebration, has Iago exit directly below it at the end of Act 2, scene 3 after the line "Dull not device by coldness and delay," pans to reveal it immediately after Iago murders Roderigo, and cuts to a shot of it at the moment Desdemona dies. The moments Welles chooses are significant. The first two occur immediately before Iago takes a major step in his attack: i.e., before the tricking of Cassio and before the "temptation" scene (Welles places the "temptation" scene directly after Act 2, scene 3). The last two instances take place immediately after we witness the tragic consequences of his deeds in the deaths of Roderigo and Desdemona. Thus we are reminded of Iago's ultimate punishment both in the preparation and in the culmination of his evil machinations. Of course, there is a neat symmetry to Welles's choices as well: the attempt to destroy Cassio ultimately leads to the death of Roderigo (shots one and three), whereas the manipulations of the "temptation" scene culminate in the death of Desdemona (shots two and four). But what is most important to stress here is that the director exposes an aspect of time that even Iago cannot control. By opening his film with the visuals of the final moment and by demonstrating Iago's eventual doom with recurring shots of his cage, Welles suggests that even though Iago works in time to destroy time, "the old arbitrator time" eventually triumphs and exposes his treachery.

CONCLUSION

New Relationships

> . . . it appears that we could never find our way about in this world if we were not . . . attuned to relationships.[1]

I

The strength of a performance-centered criticism for Shakespearean drama lies in the kinds of relationships it uncovers, relationships that will always be slightly different with every new production. Advocates of this critical orientation continually stress the importance of seeing the plays on the stage they were intended for; after all, Shakespeare wrote these plays, not to be read, but to be seen and heard. But the issue of intention pales when one realizes that the potency of performance criticism is its capacity to explore how a given production organizes the material of Shakespearean drama into new relationships. An exploration of the plays in performance unveils the relationship of visual and aural material, of character and setting, of spoken language and physical gesture, of a whisper and a tear—all of which function according to the larger relationship between the time and space attributes of production. With every new production, with every new conceptual focus, those relationships change, new contexts emerge, and the elements in Shakespeare's drama unfold in endless permutations that have kept the plays alive for four centuries. Moreover, the plays achieve meaning for the spectator through these relationships; to engage as a spectator with a performance of Shakespearean drama is to work with the relationships the production offers, to learn from elements it juxtaposes, to imaginatively organize the performance event. The "virtual dimension" (itself built on a relationship) of the Shakespeare play is a dimension constituted by the perceiving subject who works with the elements of the performance. To adapt Gombrich's famous argument, performance criticism must examine, not only the specific relationships offered in the context of production, but the nature of our reactions to those relationships as well.[2]

Performance criticism must build its foundation on psychological theories

145

of "reading" art as much as it does on the semiological systems of performance (especially with respect to questions of time and space) and the more
phenomenological and psychoanalytic questions of the subject–object relationship. The work of Meyerhold, Brecht, Grotowski, and Barba combine
with the Russian Formalists, Gombrich, Iser, Pavis, and contemporary
critics like Bert States, to construct the basis for a rigorous performance
criticism. The insights of all these figures, whether tacit or expressed in my
argument, inform this study. We may now more precisely define performance criticism as that approach that examines how the text functions as a
product of the relationships created through production and the way the
spectator, in turn, organizes that material.

Different kinds of productions, however, create different kinds of relationships. A study of Shakespeare's plays on stage is an examination of the
drama as a product of the distinctive potentials of live presentation. The
film medium, however, operates according to different laws and therefore
organizes the material of Shakespearean drama in a different manner. Film
is obviously not the production context "intended" by Shakespeare (neither,
for that matter, is the modern stage), but as a performance medium in its
own right, the cinema creates specific kinds of relationships among the
elements of the drama to activate a peculiar imaginative response. What we
can learn about Shakespeare through productions on the screen results
precisely from the new relationships they offer. Film can be a critical tool. I
have thus organized this study according to the relationships I see as most
pertinent to establishing a performance criticism of Shakespeare through
film adaptations, an approach ultimately informed by that more comprehensive "super-relationship" of cinematic time and space.

II

Some of the most important new relationships that come from Shakespeare
on film are a result of the multiple perspectives that the cinematic medium
offers. Visually, the films create what Eisenstein calls "dynamization in
space," a multiplicity of perspectives on a given object that the mind organizes into a sensible whole. Film stimulates the imagination to work with a
world of fragments that result either from breaking up phenomena in montage, or from showing many perspectives on a more "complete" or "whole"
object of observation, or both. In the case of Shakespeare on film, the elements of the *play* function according to this dynamic, and the spatial field of
the drama finds definition in the perspectives that the medium offers. Performance criticism of Shakespeare films thus exposes new relationships between
subjective and objective points of view, between the alternating subjective

perspectives of two (or more) characters in a scene, between subjective states of mind and spatial relationships of the *mise-en-scène* that contrast with those subjective states. Moreover, the camera can take us on a journey among the subtextual elements of the play and give visual form to a level of meaning that stands in a new relationship with other, more ostensible, levels of that play. Add to the multiplicity of the visual field the various elements of the aural field and the permutations become even more complex.

Similarly, multiple perspectives operate in the temporal organization of the film and new relationships emerge as a result. The film works with the already fluid nature of Shakespeare's own temporal structure to form relationships both synchronously and diachronically among key elements of the drama. The spectator works with simultaneous action (through relationships of the visual field and of the visual and aural fields in juxtaposition) in a dynamic not unlike the multiple perspectives that define the spatial attributes of the medium. In addition, to state the obvious, the viewing process moves through time (these are movies, after all) and filmmakers juxtapose moments throughout the course of the film (by exploiting techniques that trigger memories and create expectations for future events) to activate a continuous process of anticipation and retrospection, a process at the heart of the temporal journey of viewing. The spectator works with the relationships produced by the temporal structure of the film to create meaning for the play on the screen.

Film also works with relationships of locale in Shakespeare's plays to create a distinctive spatial dynamic. Specifically, the juxtaposition of inside and outside settings participates in the definition of the kinds of worlds in which the dramas unfold, a definition that results not only from fundamental antitheses, but from a process of reciprocity as well. The configuration of interior and exterior worlds illuminates new relationships of the closed and the open, of movement and stasis, of intimacy and distance, with the result of exposing both the psychological and political character of the plays as well as the tensions that inform them most significantly. In a sense, interior and exterior worlds function in a relationship similar to what Frye has shown us to be the dynamic of worlds in Shakespeare's romantic comedies; indeed, one can understand inside and outside as mutual "greenworlds," as worlds that work together through processes both of collision and harmonious balance, combining to articulate and define the larger space in which the drama unfolds. The desolate exterior of Kozintsev's *Lear* emerges in the film as a kind of ironic greenworld, a barren locale for growth and understanding.

Part of the magic of the greenworld of the comedies is, like film and Shakespearean drama, a result of the new relationships that come about from new contexts. The film medium can provide events of the drama, like the storm in *King Lear*, with a special context for enactment and therefore lead to the building of new relationships between action and environment. Film provides a new space for Shakespeare's poetry, a space in which an event can find new life because it appears in a fresh context. Similarly, the close-up provides a spatial context for the events of the drama to unfold—a private space, an intimate space, one that can serve as the place for covert action, for the most intimate expressions of emotion, and as a locale for a tortured psyche. The close-up is a shot packed, not only with the secret and hidden elements of the drama, but with its social character as well; by exposing, in powerful detail, the complex attitudes of Shakespeare's figures in a given scene in the play, the close-up becomes the field that can articulate the social *gest* of the performance moment. Moreover, in juxtaposition with the long shot, the close-up can build relationships of size, of giants and dwarfs, and can give form to our own ambivalence towards the playwright's figures of vice. Finally, just as Shakespeare informs his work with a theatrical self-consciousness through relationships between content and form, so, too, can the filmmakers of cinematic adaptations create a similar filmic self-consciousness. The new relationships evident here are built not so much on elements in the film itself as they are on the interactions of spectator and performance. Shakespeare teaches us about the act of viewing in the theater and the creative endeavors of the art; film, by pulling us in and out of theatrical and filmic spaces, makes us aware of the nature of cinematic adaptation and its capacity to realize a stage play on the screen.

Finally, the new relationships that emerge from cinematic adaptations shed light on Shakespeare's dramaturgical design, and the films stand as important contributions to the critical histories of the plays. In a sense, as a process of defamiliarization, cinematic adaptation guides us to ask different questions about the drama, to explore the plays in a distinctive way because they appear to us in a new context. Welles's *Othello* is a particularly convincing example of cinema's potential to open fresh interpretive questions, but all the films examined in this study encourage the same process on varying levels.

At the heart of film's ability to serve as critical tool is the formation of relationships built, not only on the dynamics of the drama, but also on the spectator's imaginative response to those dynamics. Shakespeare films have the remarkable capacity to give form to, say, the manipulations of a vice-figure, such as Richard III, and our corresponding responses to those machinations. In patterns of close and long shots, the visuals imitate our very ambivalence towards Richard III; at the same time, the range and distance of the shots parallel his own techniques of achieving control. Richard knows when to push and when to hold back, when to ensnare us in close-up and when to give us a little breathing room in the long shot. The vice knows that without the distance to complement his more overt attack (or, in cinematic terms, without the relaxation of the "long shot" to strengthen the "close-up" of his assault), his powers would be nothing. The close-up, likewise, can give form to the prison of the mind and provide a spatial field with characteristics that, in and of themselves, articulate subjective experience. Similarly, the familiar shot–counter-shot pattern employed by Welles in the new King's rejection of Falstaff gives form to the ambiguities and tensions of Shakespeare's design as well as to our ambivalence that results from that design. As Iago works with time to destroy time for Othello, the film, with its resources, realizes that very process; form and content are one as Welles creates the very chaos in the visual and aural fields of cinematic performance that is at the heart of Othello's demise.

But the chaos Welles creates in his film as the equivalent of Iago's weapon and Othello's loss of self achieves that equivalence only with a spectator who can work with the new relationships the filmmaker offers. Gombrich celebrates this creative human capacity when he suggests that "there is perhaps no limit to the systems of forms that can be made the instrument of artistic expression in terms of equivalence."[3] Film's representation of Shakespearean drama operates through a layering of various artists (playwright, filmmaker, actor, designer, photographer, etc.) in relationship to the perceiving subject, the "beholder," to use Gombrich's term, who, with "Pygmalion's power," meets the film halfway, and, in the instance of Welles's *Othello*, gives to moving pictures in chaotic design a specificity that he or she reads as temporal disorientation. The cinema, as a vehicle for producing Shakespeare's plays, enables us to see that our capacity as creative participants in the drama expands and changes with the new medium—we learn something about ourselves as we engage in the viewing process. And through that process, we are still in movement.

Notes

Introduction

1. Robert Hamilton Ball, *Shakespeare on Silent Film* (New York: Theatre Arts Books, 1968); Roger Manvell, *Shakespeare and the Film* (New York: A.S. Barnes, 1971); Jack Jorgens, *Shakespeare on Film* (Bloomington: Indiana University Press, 1977); Charles W. Eckert, ed. *Focus on Shakespearean Films* (Englewood Cliffs, N.J.: Prentice-Hall, 1972).

2. After completion of this book, two full-length studies were published: Bernice W. Kliman, *Hamlet: Film, Television and Audio Performance* (Cranbury, N.J.: Fairleigh Dickinson University Press, 1988); and Anthony Davies, *Filming Shakespeare's Plays* (New York: Cambridge University Press, 1988). Unfortunately, both studies appeared too late to consider carefully, and, consequently, neither informs the findings of my work.

3. *Shakespeare on Film Newsletter,* eds. Bernice W. Kliman and Kenneth S. Rothwell (University of Vermont, 1976–).

4. See, for example, *Shakespeare Survey,* Vol. 39, "Shakespeare on Film and Television" (Cambridge University Press, 1987); and *Shakespeare on Film and Television: An Anthology of Essays and Reviews,* eds. J. C. Bulman and H. R. Coursen (Hanover, N.H.: University Press of New England, 1988).

5. Dudley Andrew provides a comprehensive bibliography of film theory in *Concepts in Film Theory* (New York: Oxford University Press, 1984). In addition, I recommend the first-rate collection of essays edited by Philip Rosen entitled *Narrative, Apparatus, Ideology* (New York: Columbia University Press, 1986). Rosen's anthology contains some of the finest and most penetrating studies on film that I have read to date.

6. Martin Heidegger, "The Question Concerning Technology," in *The Question Concerning Technology, and Other Essays,* trans. William Lovitt (New York: Harper & Row, 1977), 16.

7. See especially Jack Jorgens, *Shakespeare on Film,* and Roger Manvell, *Shakespeare and the Film.*

8. *Macbeth*—Mercury Productions, released by Republic Pictures. Producer, director, and screenplay: Orson Welles. Photographer: John L. Russell and William Bradford. Art director: Fred Ritter. Editor: Louis Lindsay. Music: Jacques Ibert, conducted by Efrem Kurtz. Sound: John Stransky, Jr., and Gary Harris.

9. *Othello*—Mercury Productions, released by United Artists. Producer and director: Orson Welles. Photographers: Anchise Brizzi, G. Araldo, George Fanto. Editors: Jean Sachs, Renzo Luoidi, John Shepridge. Music: Francesco Lavagnino, Alberto Barberis, conducted by Willi Ferrero.

10. *Chimes at Midnight*—International Films Espagnola Alpine Productions (Spain/ Switzerland), released in U.S. by Peppercorn-Wormser, Inc., U-M Film Distributors. Producers: Emiliano Piedra and Angel Escoloano. Executive producer: Alessandro Tasca. Director and screenwriter: Orson Welles. Photographer: Edmond Richard. Art directors: Jose Antonio de la Guerra and Mariano Erdorza. Editor: Fritz Muller. Music: Angelo Francesco Lavagnino, conducted by Pierluigi Urbini.

11. Quoted in Manvell, *Shakespeare and the Film*, 59.

12. For an eloquent account of the erratic nature of the making of this film, see Michael MacLiammoir, *Put Money in Thy Purse* (London: Methuen, 1952). MacLiammoir played the part of Iago.

13. Welles's deep admiration for Verdi is evident from the fact that he followed the composer's route in the making of his Shakespeare films, creating, in the same sequence as the operas, adaptations of *Macbeth, Othello* and *Falstaff*.

14. For an account of the project, see Peter Noble, *The Fabulous Orson Welles* (London: Hutchinson, 1956), 103.

15. Manvell, *Shakespeare and the Film*, 64–71.

16. *Hamlet* (*Gamlet*)—Lenfilm, released by Lopert Pictures, 1964. Sovscope. Screenplay: Grigory Kozintsev, from the Russian-language translation by Boris Pasternak. Photographer: J. Gritsyus. Art director: E. Ene and G. Kropachev. Editor: E. Makhankova. Music: Dimitri Shostakovich, played by Leningrad Philharmonic Orchestra, conducted by N. Rabinovitch. Sound: B. Khutoryanski. In Russian with English subtitles.

17. *King Lear* (*Korol' Lir*)—Lenfilm, released by Artkino Pictures, 1970. Sovscope. Screen adaptation: Grigory Kozintsev, from the Russian-language translation by Boris Pasternak. Photographer: Jonas Gritsyus. Design: Evgeni Ene, Ulitka, S. Virsaladze. Music: Dimitri Shostakovich. Sound: E. Vanuts. In Russian with English subtitles.

18. Trans. Joyce Vining (New York: Hill and Wang, 1966).

19. Trans. Mary Mackintosh (Berkeley and Los Angeles: University of California Press, 1977). Originally published in Russian, 1973.

20. *Time and Conscience*, 156.

21. Quoted in *Meyerhold on Theatre*, ed. Edward Braun (New York: Hill and Wang, 1969), 190.

22. *Time and Conscience*, 24.

23. Ibid., 32, 40.

24. Ibid., "Introductory Note." The original title of *Time and Conscience* in Russian would translate literally as *Our Contemporary: William Shakespeare*, a title that duplicates the one that Jan Kott gave to his study: *Shakespeare: Our Contemporary*, trans. Boleslaw Taborski (New York: Doubleday and Company, Inc., 1966). Kozintsev changed the title to *Time and Conscience* for the English translation to avoid confusion with Kott's work.

25. *Henry V*—A Two Cities Film, released by United Artists. Producer and director: Laurence Olivier. Screenplay: Alan Dent and Laurence Olivier. Photographer: Robert Krasker. Art director: Paul Sheriff. Editor: Reginald Beck. Music: William Walton, played by the London Symphony, conducted by Muir Mathieson. Sound: John Dennis and Desmond Dew.

26. *Hamlet*—A Two Cities Film, released by J. Arthur Rank Organisation. Producer and director: Laurence Olivier. Photographer: Desmond Dickenson. Art director: Cameron Dillon. Editor: Helga Cranston. Music: William Walton, played by

the Philharmonia Orchestra, conducted by Muir Mathieson and John Hollingsworth. Sound: Jon W. Mitchell and Harry Miller.

27. *Richard III*—London Films, released by Lopert Films Distributing Corporation. VistaVision. Producer and director: Laurence Olivier. Photographer: Otto Heller. Art director: Carmen Dillon. Editor: Helga Cranston. Music: William Walton.

28. For the most comprehensive study on Olivier's Shakespeare films, see Dale Silviria, *Laurence Olivier and the Art of Film Making* (Cranbury, N.J.: Fairleigh Dickinson University Press, 1985).

29. *King Lear*—Filmways, Inc., in association with the Royal Shakespeare Company, released by Columbia Pictures. Producer: Michael Birkett. Director and screenplay: Peter Brook. Photographer: Henry Kristiansen. Editor: Kasper Schyberg.

30. *Macbeth*—Playboy Productions/Caliban Films, released by Columbia Pictures. Producer: Andrew Braunsberg. Screenplay: Roman Polanski and Kenneth Tynan. Photographer: Gil Taylor. Editor: Alastair McIntyre. Music: The Third Ear Band. Choreographer: Sally Gilpin.

31. The film and Brook's stage production that precedes it were influenced by Jan Kott's essay *"King Lear* or *Endgame"* in *Shakespeare Our Contemporary*, an essay pointing out what is to Kott the cold, unforgiving worlds created by both Shakespeare and Beckett and the nihilism at the foundation of both plays.

32. *Hamlet*—Woodfall Films–Filmways Ltd., released by Columbia Pictures. Producer: Neal Hartley. Executive producers: Leslie Linder and Martin Ransohoff. Photographer: Gerry Fisher. Design: Jocelyn Herbert. Editor: Charles Rees. Music: Patrick Gowers. Sound editor: Don Deacon. The film is an adaptation of a stage production Richardson directed "in the round" in London's Round House Theatre.

33. See "The Work of Art in the Age of Mechanical Reproduction," in *Illuminations*, trans. Harry Zohn, ed. Hannah Arendt (New York: Schocken Books, 1969), 236.

Chapter 1: Spatial Multiplicity

1. *Rococo to Cubism in Art and Literature* (New York: Random House, 1960), 264. Sypher argues convincingly that the cinema, like the cubist artifact, operates on the basis of a multiplicity of perspectives and organizes the spectator's experience in terms of its constantly moving "plural relations." His point is pertinent to the central question of this chapter, which addresses how the Shakespeare text is organized by the multiplicity of cinematic space.

2. "The Reading Process: A Phenomenological Approach," in *The Implied Reader* (Baltimore: Johns Hopkins University Press, 1974), 279. Cf. Roman Ingarden, *Das literarische Kunstwerk* (Tübingen, 1960), 270 ff.

3. "Anticipation and retrospection" is the phrase Iser offers when trying to define the specific nature of the temporal dimension of the reading process. See Chapter 6 for a more detailed discussion of Iser's concept and its significance to cinematic time.

4. The distinction between "seeing" and "recognizing" is the one Viktor Shklovsky offers in "The Resurrection of the Word," his seminal essay on defamiliarization (1914). See *Russian Formalism: A Collection of Articles and Texts in Translation*, ed. Stephen Bann and John E. Bowlt (Edinburgh: Scottish Academic Press, 1973).

5. See Gerald Else's exploration of the issue in *Aristotle's Poetics: The Argument* (Cambridge: Harvard University Press, 1963). Else argues convincingly that, for

Aristotle, identification has to do specifically with the *action* of the tragedy, with an understanding of the "mistake" made by the hero, a mistake in the identity of blood kin. The spectator "identifies" in the sense that he or she, too, under similar conditions, could make the same mistake. The commonly accepted notion of "sympathy" for the individual character, for "one like ourselves" (and the concomitant sense that pity and fear are the direct result of this sympathy), is more a product of Renaissance interpreters like Castlevetro and the later neoclassicists. G. E. Lessing, in the *Hamburg Dramaturgy*, stresses the notion of identification to the extent of defining the process of "fear" in tragedy as a result of a "compassion referred back to ourselves." See the translation by Helen Zimmern (New York: Dover Publications, 1962).

6. For a similar definition of "identification" as part of the reading process, see Iser, "Reading Process," 291.

7. Interestingly, Welles's first Shakespeare film, though problematic in many ways, is a particularly rich place to examine this imaginative paradigm. Despite the hasty shooting period (and the accompanying technical disasters), his conceptual reductionism (*Macbeth* as morality play), and his mediocre cast, Welles still offers a film one can use as a centerpiece of a study of the spatial multiplicity of filmed Shakespeare.

8. *Shakespeare Our Contemporary*, 96.

9. David Bevington, ed., *The Complete Works of Shakespeare*, 3rd ed. (Glenview, Illinois: Scott, Foresman, 1980). All quotations for all plays I discuss in the study are taken from this collection.

10. *Shakespeare and the Allegory of Evil: The History of a Metaphor in Relation to His Major Villains* (New York: Columbia University Press, 1958).

11. Critics repeatedly quote Olivier's contention that, despite one's instinct for a close-up at the climactic moments of a soliloquy, they are actually moments that should be shot at a distance. He saw as embarrassing Cukor's tight shot of Norma Shearer in the climax of the potion scene in *Romeo and Juliet* (See Manvell, *Shakespeare and the Film*, 37–38). In his *Henry V*, he repeatedly used close shots for the opening of key soliloquies and then moved to a long shot for the climax. "This is a real actor's directing solution," says Peter Brook. While I believe that the technique is generally a key to Olivier's directorial strategies, I do not recognize it as particularly significant to the moment of *Richard III* that I address in this chapter.

12. I can imagine an instance in which the reverse would be true: with a close-up we could feel repelled, while the distance of a long shot could give us the "space" and freedom of enticement and make us want to reach for what appears distant. I do not believe, however, that Olivier's film operates this way. Richard is attractive in close-up. Indeed, we understand something of what might be irresistible to Anne later on as Richard woos her in close range.

13. *3 Henry VI*, (3.2.153). Olivier, in a move that dates back to a fairly standard nineteenth-century practice, combines Gloucester's soliloquies from *3 Henry VI* (3.2.124–195) and *Richard III* (1.1.1–41) at the opening of the film.

14. For a detailed history on this point, see Marvin Rosenberg, *The Masks of Macbeth* (Berkeley, University of California Press, 1978), 441 ff.

15. Greenblatt, *Shakespearean Negotiations* (Berkeley: University of California Press, 1988), 63.

16. Sympathy for Falstaff was evident as early as 1709, when Nicholas Rowe asked "whether some people have not in remembrance of the diversion he [Falstaff] afforded 'em, been sorry to see his friend use him so scurvily when he comes to the

crown in the end of the Second Part of *Henry IV.*" Quoted in Moody E. Prior, "Comic Theory and the Rejection of Falstaff," *Shakespeare Studies*, no. 9 (1976), 159. See also the famous defense of Falstaff by Maurice Morgann in "An Essay on the Dramatic Character of Sir John Falstaff" (1777) in *Eighteenth Century Essays on Shakespeare*, ed. Nichol Smith, 2nd. ed. (Oxford: Oxford University Press, 1963), 218–31. For a similar perspective, see A. C. Bradley, "The Rejection of Falstaff" (1902), in his *Oxford Lectures on Poetry* (Bloomington: Indiana University Press, 1961); and W. H. Auden, "The Prince's Dog," in his *The Dyer's Hand* (New York: Random House, 1948).

17. An entire group of critics on Falstaff holds that the rejection is not only necessary but a dramatically sound and welcome conclusion. See J. Dover Wilson, *The Fortunes of Falstaff* (Cambridge, 1943; reprint New York: Cambridge University Press, 1979), 32. C. L. Barber sees Falstaff as the Lord of Misrule who reigns temporarily but must ultimately fall at the end of the saturnalian festivities. See *Shakespeare's Festive Comedy: A Study of Dramatic Form and Its Relation to Social Custom* (Princeton: Princeton University Press, 1959), 192–221. J. I. M. Stewart and Philip Williams see the Knight in psychoanalytic terms; see J. I. M. Stewart, *Character and Motive in Shakespeare* (New York: Longmans & Green, 1949); and Phillip Williams, "The Birth and Death of Falstaff Reconsidered," *Shakespeare Quarterly*, Vol. 8 (1957), 359–65. There are also those who see the rejection in political terms; see, for example, Stephen Greenblatt's essay, "Invisible Bullets," in *Shakespearean Negotiations* (Berkeley: University of California Press, 1988), 21–65; and William Empson, "Falstaff and Mr. Dover Wilson," *The Kenyon Review*, no. 15 (1953), 222. For a Biblical parallel of the Last Judgment, see Edward I. Berry, "The Rejection Scene in *2 Henry IV*," *Studies in English Literature, 1500–1900*, Vol. 17 (1977), 201–21. Bernard Spivack discusses the rejection in light of the traditional banishment of the vice-figure of the morality play. See "Falstaff and the Psychomachia," *Shakespeare Quarterly*, Vol. 8 (1957), 449–59.

18. Welles's Prince Hal is not a figure who demonstrates Machiavellian prudence in his exercise of power; he develops throughout the film into a king who assumes power after a number of important stages that Welles is at pains to make evident. For a thorough treatment of this development, see Samuel Crowl, "The Long Goodbye: Welles and Falstaff," *Shakespeare Quarterly*, Vol. 31 (1980). The spectator of *Chimes* is aware that Hal grows into kingship in this film as much as she or he is aware of the filmmaker's glorification of and identification with Falstaff. The rejection scene on film, therefore, culminates a tension of interpretive strategies present throughout the film.

19. There are, finally, critics who applaud the rejection and agree with Samuel Johnson that "the fat knight has . . . nothing in him that can be esteemed. . . ." See *Samuel Johnson on Shakespeare*, ed. W. K. Wimsatt, Jr. (New York: Hill and Wang, 1960), 88. See also, for the same line of argumentation, S. T. Coleridge, *Coleridge's Writings on Shakespeare*, ed. Terence Hawkes (New York: Capricorn Books, 1959), 246; E. E. Stoll, *Shakespeare Studies: Historical and Comparative in Method* (New York: Macmillan, 1927).

20. Greenblatt, *Negotiations*, 37.

21. "A Short Organum for the Theatre," in *Brecht on Theatre*, trans. John Willett (New York: Hill and Wang, 1964), 197.

22. See Chapter 2, where I treat in more detail the nature of the filmmaker's conceptual focus.

23. *Shakespeare on Film*, 152.

24. *Shakespeare's Imagery and What It Tells Us* (New York: Cambridge University Press, 1935), p. 326.

Chapter 2: Inside-Out

1. *The Poetics of Space*, trans. Maria Jolas (Boston: Beacon Press, 1969), 211–31.

2. Ibid., 215.

3. Jorgens, *Shakespeare on Film*, 246.

4. *Time and Conscience*, 65.

5. *Poetics of Space*, 215.

6. James Naremore, *The Magic World of Orson Welles* (New York: Oxford University Press, 1978), 265.

7. Crowl, "Long Goodbye," 374.

8. Though Welles locates all of the scenes at the Boar's Head inside, he does, significantly, shoot Hal's first soliloquy ("I know you all") just outside the tavern building (with Falstaff nearby), where we are able to see, in the distance, Henry's castle. In time, Hal will bring England out of its protective shell, though at this point one still detects an unresolved tension. As Samuel Crowl points out, "In Welles's strategic use of topography and spatial relationships, Hal here stands caught between the tavern and the castle, between Falstaff's inviting smile and the bleak landscape leading toward the fortifications of responsibility" (p. 375).

9. "Welles and Falstaff: An Interview by Juan Cobos and Miguel Rubio," *Sight and Sound*, Vol. 35 (1966), 161.

10. Dover Wilson, *Fortunes of Falstaff*, 32; Empson, "Falstaff and Mr. Dover Wilson," 222.

11. *Time and Conscience*, 224.

12. "*Hamlet* and *King Lear*," in *Shakespeare 1971*, ed. Clifford Leach and J. M. R. Margeson (Toronto: University of Toronto Press, 1972), 192.

13. Jorgens, *Shakespeare on Film*, 227.

14. *Time and Conscience*, 166.

15. For an excellent analysis of how the first scene of *Hamlet* places the Ghost in a Christian context, see Eleanor Prosser's *Hamlet and Revenge*, 2nd ed. (Stanford: Stanford University Press, 1971), 97–117.

16. *Time and Conscience*, 154.

17. Ibid., 152.

18. Ibid., 168.

19. Ibid., 250.

20. *Space of Tragedy*, 115.

21. *Time and Conscience*, 163.

22. Ibid., 168–69.

23. Patrice Pavis, *Languages of the Stage: Essays in the Semiology of the Theatre* (New York: Performing Arts Journal Publications, 1982), 138.

24. *Questions of Cinema* (London: MacMillan, 1981), 53.

Chapter 3: Houseless Heads

1. "On the Tragedies of Shakespeare," in *Essays* (Folio Society, 1963), 32.

2. *Shakespearean Tragedy*, 2nd. ed. (1905; reprint London: Macmillan, 1963), 221–22.

3. Maynard Mack, *King Lear in Our Time* (Berkeley: University of California Press, 1972), 36; Marvin Rosenberg, *The Masks of King Lear* (Berkeley: University of California Press, 1972), 79.

4. *Space of Tragedy*, 227.

5. Bradley, *Shakespearean Tragedy*, 221.

6. *Space of Tragedy*, 231–32.

7. Ibid., 229.

8. "*King Lear:* Draft Shooting Script," (London: Ondanti Script Services, 1968). From a microfilm copy (Film Acc. 428.3) deposited in the Folger Shakespeare Library, Washington, D.C.

9. *Great Reckonings in Little Rooms: On the Phenomenology of Theater* (Berkeley: University of California Press, 1985), 57, 58.

Chapter 4: Expanding Secrets

1. Lillian Gish, *The Movies, Mr. Griffith, and Me* (London: W. H. Allen, 1969), 59–60.

2. Bela Balazs, *Theory of the Film: Character and Growth of a New Art*, trans. Edith Bone (New York: Dover Publications, 1970), 55.

3. "The Work of Art in the Age of Mechanical Reproduction," 236–37.

4. *Theory of the Film*, 60.

5. See Meyerhold's essay "The Stylized Theatre" in *Meyerhold on Theatre*, 58–64. "Ultimately the stylistic method presupposes the existence of a fourth *creator* in addition to the author, the director and the actor—namely, the spectator . . . [who] *fill[s] in* those details *suggested* by the stage action." See also E. H. Gombrich's chapters on "the beholder's share" in *Art and Illusion: A Study in the Psychology of Pictorial Representation* (Princeton: Princeton University Press, 1972).

6. Balazs, *Theory of the Film*, 61.

7. As Michael Mullin points out, Richardson's cinematic technique parallels his stage direction: "As his staging used theatre in the round to break down the convention of isolated actors on a picture stage, Richardson's camerawork used close-ups to break down the conventional separation of audience and actors, allowing us to look closely at each actor's features, putting us in their midst, so that we see the play as the actors do. We are 'onstage' with the others." See "Tony Richardson's *Hamlet:* Script and Screen," in *Literature/Film Quarterly*, Vol. IV, no. 2 (Spring, 1976), 123–31.

8. *What is Cinema?*, trans. Hugh Gray (Berkeley: University of California Press, 1971), 110.

9. *The Acoustic Mirror: The Female Voice in Psychoanalysis and Cinema* (Bloomington: Indiana University Press, 1988), 53.

10. Polanski and his co-writer, Kenneth Tynan, interpolated these lines from earlier in the scene, lines that, in the original text, Macbeth says while the witches are still present; it is a passage in which he exhorts them to tell of their powers.

11. For a critique of Polanski's decision to show the murder, a point of much controversy surrounding the film, see Robert Ornstein's article "Interpreting Shakespeare: The Dramatic Text and the Film," *The University of Dayton Review*, Vol. 14, no. 1 (1979–80), 55–61. Ornstein argues that the filmmaker's strategy destroys the balance of Shakespeare's play, dominates the imagination with visuals that eclipse the focus on Lady Macbeth while the deed is done, and makes pallid the following scene between them.

12. In the text, Shakespeare calls for the corpse of Henry VI, Anne's father-in-

law. Olivier's change emphasizes the irony of the murderer Richard's vying for the hand of the widow of his victim and downplays Shakespeare's specific political point in the original. The political impact of Olivier's film exists more in the relationship between Anne and Gloucester than it does in the significance of the moment in England's history.

13. *Brecht on Theatre*, 198.

14. *Brecht: The Man and His Work* (New York: Doubleday, 1961), 134.

15. Roland Barthes, "Seven Photo Models of *Mother Courage*," *The Drama Review*, Vol. 12, no. 1 (Fall 1967), 45. Barthes looks at photographs of *Mother Courage* that to him contain the basic *gest* of a given moment and illuminate the political and social meaning of the play. Barthes's work with the photographic still in this article as well as in "The Third Meaning," in *Image-Music-Text*, trans. Stephen Heath (New York: Hill and Wang, 1977), 52–68, informs my approach to the following study of the *gestic* content of the close-up.

16. Roland Barthes, "Third Meaning," 67.

Chapter 5: Local Habitations

1. For the most thorough studies on metadrama in Shakespeare (a topic that has received much critical attention), see the following: James Calderwood, *Shakespearean Metadrama* (University of Minnesota Press, 1971); and idem, *Metadrama in Shakespeare's Henriad* (Berkeley: University of California Press, 1979); Anthony B. Dawson, *Indirections: Shakespeare and the Art of Illusion* (Toronto: University of Toronto Press, 1978); Philip Edwards, *Shakespeare and the Confines of Art* (London: Methuen, 1968); Robert Egan, *Drama within Drama: Shakespeare's Sense of His Art* (New York: Columbia University Press, 1975); Kirby Farrell, *Shakespeare's Creation: The Language of Magic and Play* (Amherst: University of Massachusetts Press, 1975); Richard Fly, *Shakespeare's Mediated World* (Amherst: University of Massachusetts Press, 1976); Sidney Homan, *When the Theatre Turns to Itself: The Aesthetic Metaphor in Shakespeare* (Lewisburg: Bucknell University Press, 1981); Alvin B. Kernan, *The Playwright as Magician: Shakespeare's Image of the Poet in the English Public Theatre* (New Haven: Yale University Press, 1979); Lachlan Mackinnon, *Shakespeare the Aesthete: An Exploration of Literary Theory* (London: Macmillan, 1988); Philip C. McGuire and David A. Samuelson, eds., *Shakespeare: The Theatrical Dimension* (New York: AMS Press, 1979); Anne Righter, *Shakespeare and the Idea of the Play* (London: Chatto and Windus, 1964).

2. *Shakespeare on Film*, 130.

3. For the debate on the stylistic nature of the battle scenes, see the following: James Agee, *Time*, April 8, 1946, pp. 56–60; and Siegfried Kracauer, *Theory of Film* (New York: Oxford University Press, 1971), 227, 260. Both critics offer representative arguments for those who insist that the Battle of Agincourt is not realistic. Opposing this, and representative of the other side, is George W. Linden in his work *Reflections on the Screen* (Belmont, California: Wadsworth Publishing Company, 1970), 22. Bridging the two extremes is Jack Jorgens, who claims, with refreshing balance, that while the battle scenes represent "a movement toward film realism," Olivier "holds that realism in check by shifting from a theatrical style to Eisensteinian montage." *Shakespeare on Film*, 131.

4. Bosley Crowther, *The Great Films: Fifty Years of Motion Pictures* (New York: G.P. Putnam's Sons, 1967), 165–68.

5. Constance Brown points out that what we do hear is the recitation of Psalm 51 in Latin, chanted by two monks sitting near the King: " 'Against thee, thee only have I sinned . . . and thou mayst be clear when thou judgest; behold I was shapen in iniquity.' " See "Olivier's *Richard III*: A Reevaluation," in Eckert, *Focus on Shakespearean Films*, 134.

6. The *Oxford English Dictionary*, citing sixteenth-century documents of Lyly and Shakespeare, offers the following obsolete definition of *shadow:* "Applied rhetorically to a portrait as contrasted with the original; also to an actor or a play in contrast with the reality represented." (Oxford: Oxford University Press, 1971), 531.

7. Much of what is at the core of Burke's writing is an exploration of the human being as a "symbol-using" animal, whose language constitutes action. See especially his essays in *Language as Symbolic Action* (Berkeley: University of California Press, 1966).

8. I leave open the possibility of more than one "performance" because of the layering of theatrical manipulations in the activities of Shakespeare, Olivier (a double layer of actor and filmmaker), and Richard himself.

9. See especially Jay L. Halio, "Three Filmed Hamlets," *Literature/Film Quarterly*, Vol. 1 (1973): 316–20; Mary McCarthy, "A Prince of Shreds and Patches," in Eckert, *Focus on Shakespearean Films*, 64–67; Manvell, *Shakespeare and the Film*, 40–47.

10. "Olivier's *Hamlet*: A Film-Infused Play," *Literature/Film Quarterly*, Vol. 4 (1977): 305. I would add that Olivier's *Hamlet* is as much a "play-infused film" as it is a "film-infused play" and that the dynamic of filmic and theatrical qualities operates through a careful balance of the two elements.

11. "Shakespeare on Three Screens," *Sight and Sound*, Vol. 34 (1965), 68–69.

12. Jorgens, *Shakespeare on Film*, 240. My reading of the moment in the film is very similar to the analysis Jorgens offers in his book, and I try to build on the cogent argument he makes. Jorgens does not, however, concern himself with the specific juxtaposition of filmic and theatrical space to demonstrate how Brook's filmic self-consciousness operates. His point mainly serves to demonstrate Brook's direction in "shaping the script" compared to Kozintsev's.

Chapter 6: Temporal Multiplicity

1. "Shakespeare, the Imaginary Cinema and the Pre-Cinema," in Eckert, *Focus on Shakespearean Films*, 27–36.

2. The exception is Welles's *Othello*, to which I dedicate the entire following chapter.

3. *The Implied Reader*, 278.

4. It should be noted, however, that various theorists do attempt to isolate basic units of performance, if only for the purpose of analysis. Eisenstein spoke of the individual shot, the "cell" that interacts with other "cells" in montage to create the film. Bazin speaks of the deep-focus shot and the dynamic of elements within it. Metz calls the basic unit a "sequence", which constitutes an arrangement of a "visual or auditory theme" in a syntagmatic position within the discourse of the film as a whole. Still, none of these matches the sentence, which the reader examines closely or passes quickly at will.

5. Cobos and Rubio, "Welles and Falstaff," 159.

6. *Shakespeare on Film*, 111–112.

7. See especially Peter Cowie, *A Ribbon of Dreams: The Cinema of Orson Welles*

(New York: A. S. Barnes, 1973); and Charles Higham, *The Films of Orson Welles* (Berkeley: University of California Press, 1970). Both critics are often helpful in their analyses of Welles's films, but, in the case of *Chimes*, they give the director's words on Falstaff far too much credence. These critics tend to sentimentalize the Knight, as the filmmaker himself has done, in a way that runs against the tensions of the film itself.

8. See citations in Chapter 1.

9. See Cobos and Rubio, "Welles and Falstaff," 159.

10. Crowl, "Long Goodbye," 373.

11. Ibid., 376.

12. See Chapter 2.

13. *Time and Conscience*, 250.

14. Ibid., 230.

15. Ibid., 250.

16. *Space of Tragedy*, 183.

17. Ibid., 24.

18. Ibid., 207.

19. Ibid., 250–51.

20. Ibid., 238.

Chapter 7: Naming Time

1. A version of this chapter first appeared under the title "Orson Welles's *Othello*: A Study of Time in Shakespeare's Tragedy," in *Shakespeare Survey*, Vol. 39 (1987). It is reprinted with the permission of Cambridge University Press.

2. Quoted in Peter Noble, *Fabulous Orson Welles*, 215.

3. "Orson Welles: Of Time and Loss," *Film Quarterly*, Vol. 21 (1967), 21.

4. Translated by Mark Musa (New York: St. Martin's Press, 1964). See especially Chapter 25.

5. *Shakespeare Our Contemporary*, 112.

6. *The Wheel of Fire* (London: Methuen, 1960), 107.

7. *Shakespearean Tragedy*, 155.

8. *An Approach to Shakespeare*, 2nd ed. (New York: Doubleday, 1956), 138.

9. Ibid., 138.

10. Knight, *Wheel of Fire*, 109.

11. Bradley, *Shakespearean Tragedy*, 151.

12. Traversi, *Approach to Shakespeare*, 146 (emphasis mine).

13. Kott, *Shakespeare Our Contemporary*, 123.

Conclusion

1. Gombrich, *Art and Illusion*, 49.

2. See especially Chapter 1 of *Art and Illusion*. "What a painter inquires into is not the nature of the physical world but the nature of our reactions to it": 49.

3. Ibid., 373.

Select Bibliography

Andrew, J. Dudley. *The Major Film Theories*. New York: Oxford University Press, 1976.

————. *Concepts in Film Theory*. New York: Oxford University Press, 1984.

Arnheim, Rudolph. *Film as Art*. Berkeley: University of California Press, 1971.

Auden, W. H. "The Prince's Dog." In his *The Dyer's Hand*. New York: Random House, 1948.

Bachelard, Gaston. *The Poetics of Space*. Trans. Maria Jolas. Boston: Beacon Press, 1969.

Balazs, Bela. *Theory of Film, Character and Growth of a New Art*. Trans. Edith Bone. New York: Dover Publications, 1970.

Ball, Robert Hamilton. *Shakespeare on Silent Film*. New York: Theatre Arts Books, 1968.

Barasch, Frances K. "Revisionist Art: *Macbeth* on Film." *University of Dayton Review*, Vol. 14, no. 1 (1979–80), 15–20.

Barber, C. L. *Shakespeare's Festive Comedy: A Study of Dramatic Form and Its Relation to Social Custom*. Princeton: Princeton University Press, 1959.

Barber, Lester E. "This Rough Magic: Shakespeare on Film." *Literature/Film Quarterly*, Vol. 1 (1973), 372–76.

Barker, Felix. *The Oliviers*. Philadelphia and New York: J. B. Lippincott, 1953.

Barthes, Roland. "Seven Photo Models of *Mother Courage*." *The Drama Review*, Vol. 12 (1967), 45.

————. *Image–Music–Text*. Trans. Stephen Heath. New York: Hill and Wang, 1977.

Bazin, André. *Orson Welles: A Critical View*. Trans. Jonathan Rosenbaum. New York: Harper & Row, 1978.

————. *What Is Cinema?* 2 vols. Trans. Hugh Gray. Berkeley: University of California Press, 1971.

Beckerman, Bernard. *Shakespeare at the Globe*. New York: Macmillan, 1962.

Benjamin, Walter. "The Work of Art in the Age of Mechanical Reproduction." In *Illuminations*. Trans. Harry Zohn. Ed. Hannah Arendt. New York: Schocken Books, 1969.

Berry, Edward I. "The Rejection Scene in *2 Henry IV*." *Studies in English Literature, 1500–1900*, Vol. 17 (1977), 201–18.

Bessy, Maurice. *Orson Welles*. Trans. Ciba Vaughan. New York: Crown Publishers, 1963.

Bevington, David, ed. *The Complete Works of Shakespeare*. 3rd. ed. Glenview, Illinois: Scott, Foresman, 1980.

Billard, Pierre. "Chimes at Midnight." *Sight and Sound*, Vol. 34 (1965), 64–65.

Blumenthal, J. "Macbeth into Throne of Blood." *Sight and Sound*, Vol. 34 (1965), 190–95.

Bradbrook, M. C. *Themes and Conventions of Elizabethan Tragedy*. Cambridge: Cambridge University Press, 1960.

Bradley, A. C. *Shakespearean Tragedy*. 2nd. ed. 1905; reprint London: Macmillan, 1963.

———. "The Rejection of Falstaff" (1902). In his *Oxford Lectures on Poetry*. Bloomington: Indiana University Press, 1961.

Braun, Edward, ed. *Meyerhold on Theatre*. New York: Hill and Wang, 1969.

Brecht, Bertolt. "A Short Organum for the Theatre." In *Brecht on Theatre*. Trans. John Willett. New York: Hill and Wang, 1964.

Brook, Peter. "Shakespeare on Three Screens." *Sight and Sound*, Vol. 34 (1965), 66–70.

Brown, Constance. "Olivier's *Richard III*: A Reevaluation." In *Focus on Shakespearean Films*. Ed. Charles W. Eckert. Englewood Cliffs, N.J.: Prentice-Hall, 1972.

Brown, John Russell. *Shakespeare's Plays in Performance*. London: Edward Arnold, 1966.

———, ed. *Focus on Macbeth*. London: Boston & Henley, 1982.

Buchman, Lorne M. "Orson Welles's *Othello*: A Study of Time in Shakespeare's Tragedy." *Shakespeare Survey*, Vol. 39 (1987).

Bulman, J. C., and H. R. Coursen, eds. *Shakespeare on Film and Television: An Anthology of Essays and Reviews*. Hanover, N.H.: University Press of New England, 1988.

Burke, Kenneth. *Language as Symbolic Action*. Berkeley: University of California Press, 1966.

———. *A Grammar of Motives*. Berkeley: University of California Press, 1969.

Calderwood, James. *Shakespearean Metadrama*. Minneapolis: University of Minnesota Press, 1971.

———. *Metadrama in Shakespeare's Henriad*. Berkeley: University of California Press, 1979.

Carnovsky, Morris, and Paul Berry. "On Kozintsev's *King Lear*." *The Literary Review*, Vol. 24, no. 4 (1979), 408–43.

Cavell, Stanley. *The World Viewed: Reflections on the Ontology of Film*. Cambridge: Harvard University Press, 1979.

Clay, James, and Daniel Krempel. *The Theatrical Image*. New York: McGraw-Hill, 1967.

Clayton, Thomas. "Aristotle on the Shakespearean Film; or, Damn Thee, William, Thou Art Translated." *Literature/Film Quarterly*, Vol. 2 (1974), 183–89.

Cobos, Juan, and Miguel Rubio. "Welles and Falstaff." *Sight and Sound*, Vol. 35 (1966), 158–63.

Cocteau, Jean. "Profile of Orson Welles." In *Orson Welles: A Critical View*. Trans. Jonathan Rosenbaum. New York: Harper & Row, 1978.

Cohn, Ruby. *Modern Shakespeare Offshoots*. Princeton: Princeton University Press, 1976.

Collmer, Robert G. "An Existentialist Approach to *Macbeth*." *The Personalist*, Vol. 41 (1960), 484–91.

Cottrell, John. *Laurence Olivier*. Englewood Cliffs, N.J.: Prentice-Hall, 1975.

Cowie, Peter. *A Ribbon of Dreams: The Cinema of Orson Welles*. South Brunswick and New York: A. S. Barnes, 1973.

Crowl, Samuel. "The Long Goodbye: Welles and Falstaff." *Shakespeare Quarterly*, Vol. 31 (1980), 369–80.

Crowther, Bosley. *The Great Films: Fifty Years of Motion Pictures.* New York: G. P. Putnam's Sons, 1967.

Darlington, W. A. *Laurence Olivier.* London: Morgan, Grampian Books, 1968.

Davies, Anthony. *Filming Shakespeare's Plays.* New York: Cambridge University Press, 1988.

———. "Shakespeare and the Media of Film, Radio and Television: A Retrospect." *Shakespeare Survey*, Vol. 39 (1987).

Dawson, Anthony B. *Indirections: Shakespeare and the Art of Illusion.* Toronto: University of Toronto Press, 1978.

Dean, Leonard F., ed. *Shakespeare: Modern Essays in Criticism.* New York: Oxford University Press, 1957.

Dehn, Paul. "The Filming of Shakespeare." In *Talking of Shakespeare.* Ed. John Garrett. London: Hodder & Stoughton, 1954.

De Lauretis, Teresa. *Alice Doesn't: Feminism, Semiotics, Cinema.* Bloomington: Indiana University Press, 1984.

Dollimore, Jonathan, and Alan Sinfield, eds., *Political Shakespeare: New Essays in Cultural Materialism.* Ithaca, N.Y.: Cornell University Press, 1985.

Doran, Madeleine. *Endeavors of Art: A Study of Form in Elizabethan Drama.* Madison: University of Wisconsin Press, 1954.

Dowden, Edward. *Shakespeare: A Critical Study of His Mind and Art.* 3rd ed. New York: Harper & Brothers, 1872.

Drakakis, John, ed. *Alternative Shakespeares.* New York: Routledge & Kegan Paul, 1985.

Durgnat, Raymond. *A Mirror for England: British Movies from Austerity to Affluence.* New York and Washington: Praeger Publishers, 1971.

Dworkin, Martin S. " 'Stay Illusion!' Having Words about Shakespeare on Screen." *Wascana Review*, Vol. 2 (1976), 83–93.

Eckert, Charles W., ed. *Focus on Shakespearean Films.* Englewood Cliffs, N.J.: Prentice-Hall, 1972.

Edwards, Philip. *Shakespeare and the Confines of Art.* London: Methuen, 1968.

Egan, Robert. *Drama within Drama: Shakespeare's Sense of His Art.* New York: Columbia University Press, 1975.

Eisenstein, Sergei. *Film Form.* Trans. Jay Leyda. New York: Meridian Books, 1957.

———. *The Film Sense.* Trans. Jay Leyda. New York: Meridian Books, 1957.

Else, Gerald. *Aristotle's Poetics: The Argument.* Cambridge: Harvard University Press, 1963.

Empson, William. "Falstaff and Mr. Dover Wilson." *The Kenyon Review*, no. 15 (1953), 213–62.

Esslin, Martin. *Brecht: The Man and His Work.* New York: Doubleday, 1961.

———. *The Field of Drama: How the Signs of Drama Create Meaning on Stage and Screen.* London and New York: Methuen, 1987.

Farnham, Willard. *Shakespeare's Tragic Frontier: The World of His Final Tragedies.* Oxford: Basil Blackwell & Mott, 1973.

Farrell, Kirby. *Shakespeare's Creation: The Language of Magic and Play.* Amherst: University of Massachusetts Press, 1975.

Felheim, Marvin. "Criticism and the Films of Shakespeare's Plays." *Comparative Drama*, Vol. 9 (1975), 147–55.

Fluchère, Henri. *Shakespeare and the Elizabethans*. Trans. Guy Hamilton. New York: Hill and Wang, 1956.

Fly, Richard. *Shakespeare's Mediated World*. Amherst: University of Massachusetts Press, 1976.

France, Anna Kay. *Boris Pasternak's Translations of Shakespeare*. Berkeley: University of California Press, 1978.

France, Richard. "The Voodoo *Macbeth* of Orson Welles." *Yale/Theatre*, Vol. 5, no. 3 (1974), 66–77.

Fuegi, John. "Explorations in No Man's Land: Shakespeare's Poetry as Theatrical Film." *Shakespeare Quarterly*, Vol. 23 (1972), 37–49.

Garrett, John, ed. *Talking of Shakespeare*. London: Hodder & Stoughton, 1954.

Gish, Lillian. *The Movies, Mr. Griffith, and Me*. London: W. H. Allen, 1969.

Gombrich, E. H. *Art and Illusion: A Study in the Psychology of Pictorial Representation*. Princeton: Princeton University Press, 1972.

Gottesman, Ronald, ed. *Focus on Orson Welles*. Englewood Cliffs, N.J.: Prentice-Hall, 1976.

Gourlay, Logan, ed. *Olivier*. London: Weidenfeld and Nicholson, 1973.

Granville-Barker, Harley. *Prefaces to Shakespeare*. 2 vols. Princeton: Princeton University Press, 1947.

Greenblatt, Stephen. *Shakespearean Negotiations: The Circulation of Social Energy in Renaissance England*. Berkeley: University of California Press, 1988.

Griffin, Alice. "Shakespeare through the Camera's Eye." *Shakespeare Quarterly*, Vol. 17 (1966), 383–87. See also earlier articles in *ibid.*, 4 (1953): 331–36; *ibid.*, 6 (1955): 63–66; *ibid.*, 7 (1956): 235–40.

Halio, Jay L. "Three Filmed Hamlets." *Literature/Film Quarterly*, Vol. 1 (1973), 316–20.

Hapgood, Robert. "*Chimes at Midnight* from Stage to Screen." *Shakespeare Survey*, Vol. 39 (1977).

Hawkes, Terence, ed. *Coleridge's Writings on Shakespeare*. New York: Capricorn Books, 1959.

Hazlitt, William. *Characters of Shakespeare's Plays*. In Vol. 4 of *The Complete Works of William Hazlitt*. Ed. P. P. Howe. London: J. M. Dent and Sons, 1930.

Heath, Stephen. *Questions of Cinema*. London: Macmillan, 1981.

Heidegger, Martin. *The Question Concerning Technology and Other Essays*. Trans. William Lovitt. New York: Harper & Row, 1977.

Hirsch, Foster. *Laurence Olivier*. Boston: Twayne Publishers, 1979.

Hodgdon, Barbara. "Kozintsev's *King Lear:* Filming a Tragic Poem." *Literature/Film Quarterly*, Vol. 5 (1977), 153–58.

———. "Shakespeare on Film: Taking Another Look." *The Shakespeare Newsletter*, Vol. 26 (1976), 26.

Homan, Sidney R. *When the Theatre Turns to Itself: The Aesthetic Metaphor in Shakespeare*. Lewisburg, Pa.: Bucknell University Press, 1981.

———. "Criticism for the Filmed Shakespeare." *Literature/Film Quarterly*, Vol. 5 (1977), 282–90.

Higham, Charles. *The Films of Orson Welles*. Berkeley: University of California Press, 1970.

Hurtgen, Charles. "The Operatic Character of Background Music in Film Adaptations of Shakespeare." *Shakespeare Quarterly*, Vol. 20 (1969), 53–64.

Iser, Wolfgang. *The Implied Reader*. Baltimore: The Johns Hopkins University Press, 1974.

Johnson, Samuel. *Samuel Johnson on Shakespeare.* Ed. W. K. Wimsatt, Jr. New York: Hill and Wang, 1960.

Johnson, William. "Orson Welles: Of Time and Loss." *Film Quarterly,* Vol. 21 (1967), 13–24.

Jorgens, Jack. *Shakespeare on Film.* Bloomington: Indiana University Press, 1977.

———. "Shakespeare at the Movies." *Washingtonian,* May, 1976.

———. "Image and Meaning in the Kozintsev *Hamlet.*" *Literature/Film Quarterly,* Vol. 1 (1973), 307–15.

Kael, Pauline. "Orson Welles: There Ain't No Way" (1967). In her *Kiss Kiss Bang Bang.* Boston: Little, Brown, 1968.

Kauffmann, Stanley. "Notes on Theatre and Film." In *Living Images.* New York: Harper & Row, 1975.

Kermode, Frank, ed., *Shakespeare and King Lear.* London: Macmillan, 1969.

Kernan, Alvin B. *The Playwright as Magician: Shakespeare's Image of the Poet in the English Public Theatre.* New Haven: Yale University Press, 1979.

Kitchin, Laurence. "Shakespeare on the Screen." *Shakespeare Survey,* Vol. 18 (1965), 70–74.

Kliman, Bernice W. *Hamlet: Film, Television, and Audio Performance.* Cranbury, N.J.: Fairleigh Dickinson University Press, 1988.

———. "Olivier's *Hamlet:* A Film-Infused Play." *Literature/Film Quarterly,* Vol. 4 (1977).

Knight, G. Wilson, *The Imperial Theme.* London: Methuen, 1951.

———. *Principles of Shakespearean Production.* London: Faber and Faber, 1936.

———. *The Wheel of Fire.* London: Methuen, 1960.

Knights, L. C. *Some Shakespearean Themes and an Approach to Hamlet.* Stanford: Stanford University Press, 1960.

———. *Explorations.* Middlesex: Penguin Books, 1946.

Kott, Jan. *Shakespeare Our Contemporary.* Trans. Boleslaw Taborski. New York: Doubleday, 1966.

Kott, Jan, and Mark Mirsky. "On Kozintsev's *Hamlet.*" *The Literary Review,* Vol. 24 (1979), 383–407.

Kozintsev, Grigory. *Shakespeare: Time and Conscience.* Trans. Joyce Vining. New York: Hill and Wang, 1966.

———. *King Lear: The Space of Tragedy.* Trans. Mary Mackintosh. Berkeley: University of California Press, 1977.

———. " 'Hamlet' and 'King Lear.' " In *Shakespeare 1971.* Ed. Clifford Leech and J. M. R. Margeson. Toronto: University of Toronto Press, 1972.

Kracauer, Siegfried. *Theory of Film: The Redemption of Physical Reality.* New York: Oxford University Press, 1971.

Kustow, Michael. "Hamlet." *Sight and Sound,* Vol. 33 (1964), 144–45.

Lamb, Charles. "On the Tragedies of Shakespeare." In idem, *Essays.* London: Folio Society, 1963.

Lambert, J. W. "Shakespeare and the Russian Soul." *Drama,* Vol. 126 (1977), 12–19.

Leavis, F. R. "Diabolic Intellect and the Noble Hero." In his *The Common Pursuit.* Middlesex: Penguin Books, 1952.

Lemaitre, Henri. "Shakespeare, the Imaginary Cinema and the Pre-Cinema." In *Focus on Shakespearean Film.* Ed. Charles Eckert. Englewood Cliffs, N.J.: Prentice-Hall, 1972.

Lessing, G. E. *Hamburg Dramaturgy.* Trans. Helen Zimmern. New York: Dover Publications, 1962.

Levin, Harry. *Shakespeare and the Revolution of the Times*. New York: Oxford University Press, 1976.

Linden, George W. *Reflections on the Screen*. Belmont, California: Wadsworth, 1970.

Lordkipandze, Natela. "*Hamlet* on the Screen." *Soviet Literature*, no. 9 (1964), 170–73.

McCarthy, Mary. "A Prince of Shreds and Patches." In *Focus on Shakespearean Films*. Ed. Charles W. Eckert. Englewood Cliffs, N.J.: Prentice-Hall, 1972.

McBride, Joseph. *Orson Welles: Actor and Director*. New York: Jove Publications, 1977.

———. "Chimes at Midnight." In *Focus on Orson Welles*. Ed. Ronald Gottesman. New Jersey: Prentice-Hall, 1976.

McGuire, Phillip C., and David A. Samuelson. *Shakespeare: The Theatrical Dimension*. New York: AMS Press, 1979.

Machiavelli, Niccolo. *The Prince*. Trans. Mark Musa. New York: St. Martin's Press, 1964.

Mack, Maynard. *King Lear in Our Time*. Berkeley: University of California Press, 1972.

Mackinnon, Lachlan. *Shakespeare the Aesthete: An Exploration of Literary Theory*. London: Macmillan, 1988.

MacLiammoir, Michael. *Put Money in Thy Purse*. London: Methuen, 1952.

Manvell, Roger. *Shakespeare and the Film*. South Brunswick and New York: A. S. Barnes, 1971.

Mast, Gerald, and Marshall Cohen, eds., *Film Theory and Criticism*. 2nd ed. New York: Oxford University Press, 1979.

Metz, Christian. *Film Language*. Trans. Michael Taylor. New York: Oxford University Press, 1974.

———. *The Imaginary Signifier: Psychoanalysis and the Cinema*. Trans. Celia Britton, Annwyl Williams, Ben Brewster, and Alfred Guzzetti. Bloomington: Indiana University Press, 1977.

Millard, Barbara C. "Shakespeare on Film: Towards an Audience Perceived and Perceiving." *Literature/Film Quarterly*, Vol. 5 (1977), 352–57.

Monaco, James. *How to Read a Film*. Revised edition. New York: Oxford University Press, 1981.

Morgann, Maurice. "An Essay on the Dramatic Character of Sir John Falstaff." In *Eighteenth Century Essays on Shakespeare*. 2nd ed. Ed. D. Nichol Smith. Oxford: Clarendon Press, 1963.

Morris, Peter. *Shakespeare on Film*. Ottawa: Canadian Film Institute, 1972.

Mullin, Michael. "Orson Welles' *Macbeth*: Script and Screen." In *Focus on Orson Welles*. Ed. Ronald Gottesman. New Jersey: Prentice-Hall, 1976.

———. "*Macbeth* on Film." *Literature/Film Quarterly*, Vol. 1. (1973), 332–42.

———. "Tony Richardson's *Hamlet*: Script and Screen." *Literature/Film Quarterly*, Vol. 2 (1976), 123–31.

Murry, J. Middleton. *Shakespeare*. London: The Society of Authors, 1936.

Naremore, James. "The Walking Shadow: Welles's Expressionistic *Macbeth*." *Literature/Film Quarterly*, Vol. 1 (1973), 360–66.

———. *The Magic World of Orson Welles*. New York: Oxford University Press, 1978.

Nicoll, Allardyce. *Film and Theater*. New York: Thomas Y. Crowell, 1936.

Noble, Peter. *The Fabulous Orson Welles*. London: Hutchinson, 1956.

Ornstein, Robert. *A Kingdom for a Stage*. Cambridge: Harvard University Press, 1972.

———. "Interpreting Shakespeare: The Dramatic Text and the Film." *The University of Dayton Review*, Vol. 14, no. 1 (1979–80), 55–61.

Panofsky, Erwin. "Style and Medium in the Motion Pictures." In *Film Theory and Criticism*. 2nd ed. Ed. Gerald Mast and Marshall Cohen. New York: Oxford University Press, 1979.

Parker, Barry. *The Folger Shakespeare Filmography*. Washington, D.C.: Folger Shakespeare Library, 1979.

Pavis, Patrice. *Languages of the Stage: Essays in the Semiology of Theatre*. New York: Performing Arts Journal Publications, 1982.

Pearlman, E. "*Macbeth* on Film: Politics." *Shakespeare Survey*, Vol. 39 (1987).

Prior, Moody E. "Comic Theory and the Rejection of Falstaff." *Shakespeare Studies*, Vol. 5 (1976), 159–71.

Prosser, Eleanor. *Hamlet and Revenge*. 2nd ed. Stanford, Calif.: Stanford University Press, 1971.

Rabkin, Norman. *Shakespeare and the Common Understanding*. New York: The Free Press, 1967.

————, ed. *Approaches to Shakespeare*. New York: McGraw-Hill, 1964.

Raynor, Henry. "Shakespeare Filmed." *Sight and Sound*, Vol. 22 (1952), 117–21.

Richardson, Robert. *Literature and Film*. Bloomington: Indiana University Press, 1969.

Righter, Anne. *Shakespeare and the Idea of the Play*. London: Chatto & Windus, 1964.

Roemer, Michael. "Shakespeare on Film: A Filmmaker's View." *The Shakespeare Newsletter*, Vol. 26 (1976), 26.

Rose, Mark. *Shakespearean Design*. Cambridge, Mass.: The Belknap Press of Harvard University, 1972.

Rosen, Phillip, ed. *Narrative, Apparatus, Ideology*. New York: Columbia University Press, 1986.

Rosenberg, Marvin. *The Masks of King Lear*. Berkeley: University of California Press, 1972.

————. *The Masks of Macbeth*. Berkeley: University of California Press, 1978.

————. *The Masks of Othello*. Berkeley: University of California Press, 1961.

Rosenberg, Scott. "Grigori Kozintsev's *Lear*: The Conscience of the King." *The Boston Phoenix: Arts and Entertainment*. (Feb. 21, 1984), p. 10, col. 2.

Rothwell, Kenneth S. "*King Lear* on Screen: From Metatheatre to 'Meta-cinema.' " *Shakespeare Survey*, Vol. 39 (1987).

Seng, Peter J. "Songs, Time, and the Rejection of Falstaff." *Shakespeare Survey*, Vol. 15 (1962), 31–40.

Shklovsky, Victor. "The Resurrection of the Word." In *Russian Formalism: A Collection of Articles and Texts in Translation*. Ed. Stephen Bann and John E. Bowlt. Edinburgh: Scottish Academic Press, 1973.

Silverman, Kaja. *The Subject of Semiotics*. New York: Oxford University Press, 1983.

————. *The Acoustic Mirror: The Female Voice in Psychoanalysis and Cinema*. Bloomington: Indiana University Press, 1988.

Silviria, Dale. *Laurence Olivier and the Art of Film Making*. Cranbury, N.J.: Fairleigh Dickinson University Press, 1985.

Skoller, Donald S. "Problems of Transformation in the Adaptation of Shakespeare's Tragedies from Play-Script to Cinema." Ph.D. diss., New York University, 1968.

Sontag, Susan. "Theatre and Film," In *Film Theory and Criticism*. 2nd. ed. Ed. Gerald Mast and Marshall Cohen. New York: Oxford University Press, 1979.

Spencer, Theodore. *Shakespeare and the Nature of Man*. 2nd ed. New York: Macmillan, 1961.

Spivack, Bernard. "Falstaff and the Psychomachia." *Shakespeare Quarterly*, Vol. 8 (1957), 449–59.

————. *Shakespeare and the Allegory of Evil: The History of a Metaphor in Relation to His Major Villains*. New York: Columbia University Press, 1958.

Spurgeon, Caroline. *Shakespeare's Imagery and What It Tells Us*. New York: Cambridge University Press, 1935.

States, Bert O. *Great Reckonings in Little Rooms: On the Phenomenology of Theater*. Berkeley: University of California Press, 1985.

Stewart, J. I. M. *Character and Motive in Shakespeare*. New York: Longmans & Green, 1949.

Stoll, E. E. *Shakespeare Studies: Historical and Comparative in Method*. New York: Macmillan, 1927.

————. *Art and Artifice in Shakespeare*. New York: Barnes & Noble, 1962.

Styan, J. L. *Shakespeare's Stagecraft*. Cambridge: Cambridge University Press, 1967.

————. *The Shakespeare Revolution*. Cambridge: Cambridge University Press, 1977.

Sypher, Wylie. *Rococo to Cubism in Art and Literature*. New York: Random House, 1960.

Talbot, Daniel, ed. *Film: An Anthology*. Berkeley: University of California Press, 1959.

Taylor, John Russell. "Shakespeare in Film, Radio, and Television." In *Shakespeare: A Celebration, 1564–1964*. Ed. T. J. B. Spencer. Baltimore: Penguin Books, 1964.

Tillyard, E. M. W. *Shakespeare's History Plays*. New York: Macmillan, 1946.

Toliver, Harold E. "Falstaff, the Prince, and the History Play." *Shakespeare Quarterly*, Vol. 16 (1965), 63–80.

Traversi, Derek. *An Approach to Shakespeare*. 2nd ed. New York: Doubleday, 1956.

————. *Shakespeare: From Richard II to Henry V*. Stanford, Calif.: Stanford University Press, 1957.

Ulbricht, Walt. "Orson Welles' *Macbeth*: Archetype and Symbol." *The University of Dayton Review*, Vol. 14, no. 1 (1979–80), 21–27.

Watkins, Ronald. *On Producing Shakespeare*. New York: The Citadel Press, 1965.

Welles, Orson, and Roger Hill, eds. *Macbeth*. New York: Harper & Brothers, 1941.

————. "The Third Audience." *Sight and Sound*, Vol. 23 (1954), 120–23.

Welsh, James M. "To See It Feelingly: *King Lear* through Russian Eyes." *Literature/Film Quarterly*, Vol. 4 (1976), 153–58.

Williams, Phillip. "The Birth and Death of Falstaff Reconsidered." *Shakespeare Quarterly*, Vol. 8 (1957), 359–65.

Wilson, John Dover. *The Fortunes of Falstaff*. Cambridge, 1943; reprint New York: Cambridge University Press, 1979.

————. *What Happens in Hamlet*. New York: Macmillan, 1936.

Wolfflin, Henrich. *Principles of Art History: The Problem of the Development of Style in Later Art*. Trans. M. D. Hottinger. New York: Dover Books on Art, 1932.

Wollen, Peter. *Signs and Meaning in the Cinema*. 3rd. ed. Bloomington: Indiana University Press, 1972.

Yutkevitch, Sergei. "The Conscience of the King: Kozintsev's *King Lear*." *Sight and Sound*, Vol. 40 (1971), 193–96.

Index